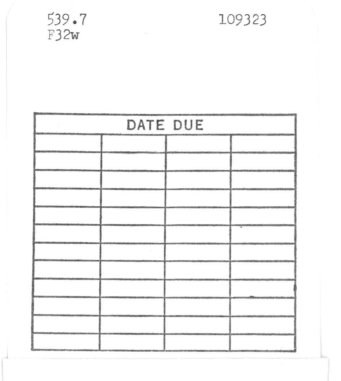

DATE DUE			

WHAT IS THE WORLD MADE OF?

Atoms, Leptons, Quarks,
and Other Tantalizing Particles

Dr. Gerald Feinberg has been Professor of Physics at Columbia University since 1965. He is the author of *The Prometheus Project*, a study of the future of mankind, as well as many articles on physics which have appeared in *Scientific American*, *The Physical Review*, *Science et Vie*, and other periodicals.

WHAT IS THE WORLD MADE OF?

Atoms, Leptons, Quarks, and Other Tantalizing Particles

Gerald Feinberg

ANCHOR PRESS/DOUBLEDAY
GARDEN CITY, NEW YORK   1977

539. 7
F 32 w
109323
ayn. 1979

Library of Congress Cataloging in Publication Data

Feinberg, Gerald, 1933–
  What is the world made of?

  Bibliography: p. 281
  Includes index.
  1. Particles (Nuclear physics) 2. Nuclear physics.
3. Atomic theory. I. Title.
QC793.24.F44   539.7
ISBN 0-385-07693-2
Library of Congress Catalog Card Number 76–18342

TO *JEREMY* AND *DOUGLAS*

Contents

Preface

I have written this book for two main reasons. One is that I find some of the discoveries of my fellow physicists to be profoundly beautiful, and I would like to share with others some of the pleasure that I get from the understanding of modern physics. I have tried here to present these discoveries in such a way that they could be understood by as many people as possible.

A second reason is that much of the research in physics over the past thirty years has been made possible by the funds provided by the U.S. government, and by the governments of other countries. I believe that we physicists have not made sufficient efforts to explain why we have undertaken the research that these funds have made possible, or to explain to nonphysicists what the outcome of this research has been. In this book, I have tried to provide such explanations to my fellow citizens who are not physicists.

Many people deserve my thanks for helping to make this book a reality. The manuscript was read by Dr. J. Bernstein, Dr. S. Morgenbesser, and Dr. E. Nagel, who made many valuable suggestions for improving it. The line drawings were done by Mr. Jim Danella. The photographs were provided me by Dr. C. Baltay and Dr. W. Y. Lee of Columbia University, Dr. P. Musset of CERN Laboratory, Mrs. Christine Nobile of Nevis Laboratory, Dr. N. Samios of Brookhaven Laboratory, Dr. Brian Thompson of the University of Rochester, by the RCA Laboratories, and by the Fermi National Accelerator Laboratory. Ms. Anne Billups did a fine job of typing the manscript from my often indecipher-

X   PREFACE

able printing. Mr. William Strachan of Anchor Books was very helpful with his editorial comments and suggestions. Finally, I thank my wife Barbara for her loving understanding while the book was being written.

New York
June, 1976

Introduction

In the twentieth century, physicists have found explanations for many phenomena that were previously known, but not understood. The precise way in which light is emitted by hot objects, the stability of atoms over millions of years or more, and the fact that some materials such as silver transmit heat and electricity much better than other materials such as rubber are examples of newly understood phenomena. In turn, the theories that explain these phenomena have also given us important new perspectives on other phenomena which formerly had been thought to be completely understood such as the composition of light and the motion of planetary bodies. In addition, physicists have been led by these theories to the discovery of qualitatively new phenomena, such as the transmutation of chemical elements and the creation and annihilation of subatomic particles.

As this progress has taken place, the very ideas which physicists use to describe nature have changed radically. Both the quantum theory and the special and general theories of relativity are examples of such new ideas. While physicists have been led to these descriptions through a combination of experiment, observation, and mathematical reasoning, and have come to accept their truth, the ideas have often seemed strange and incomprehensible to nonphysicists, even those familiar with earlier descriptions, such as Newtonian physics. Consequently, there has been little awareness among most educated nonphysicists of what physics has accomplished in this century, and what it is trying to do now.

It is perhaps inevitable that in the early stages of any scientific revolution new doctrines should appear strange to outsiders. This was the case with Newtonian physics which, when it was proposed

xii INTRODUCTION

in the seventeenth century, was rejected as unreasonable even by other leading physicists such as Gottfried Leibniz. However, more than fifty years have now passed since the quantum theory was given its definitive form in the work of Paul Dirac, surely a long enough time that novelty should no longer act as a bar to the comprehension of this theory by nonphysicists. By contrast, in the arts the ideas of fifty years ago, far from being considered too novel to comprehend, are generally thought to be old and hackneyed.

There have been two major obstacles to a more widespread understanding of modern physics. One is that the concepts introduced by physicists to explain the things they observe are often quite different from anything encountered in everyday life. A result of the theory of relativity, which states that the rate of a clock varies with the motion of the clock through space, is obviously not encountered in ordinary experience. Since it is usually easier to understand and accept the familiar than the unfamiliar, it is not surprising that nonphysicists have often found such ideas to be strange and confusing. In fact, even many physicists were slow to accept relativity theory and quantum theory when they were first proposed, largely because of the departure of these theories from ordinary experiences and from previous concepts in physics.

Nevertheless, physicists have felt constrained by their observations to formulate these new ideas and eventually to accept them. It has been the experience of several generations of physicists that once these concepts have become familiar through use, they can be manipulated as readily as the ones they replaced, and form a coherent framework into which new observations can be fitted.

A second obstacle to the public understanding of contemporary physical theories is the mathematical form in which these theories are usually expressed. However, while sophisticated mathematics is essential in the process of the discovery of new physical theories, and in the application of these theories to obtain numerical results in specific circumstances, I am convinced that a substantial comprehension of modern physics can be obtained without advanced mathematical training. There are many examples in earlier periods of books which conveyed a clear picture of some of the most difficult ideas in the physics of those periods in a way that could be understood by general readers. In this book I have not used mathematics beyond that studied in a typical secondary school course.

Obviously, it is easier to understand physics with the aid of mathematics than without it, otherwise physicists would not bother to use it. Nevertheless, if one's main interest is understanding physics, rather than creating it, it is possible and desirable to understand what one can without mathematics, rather than to abandon all efforts in the face of the thicket of formulas found in a typical physics textbook. An analogy could be made with the appreciation of music, which is possible without any knowledge of the theory of harmony, as opposed to the composition of music, which is quite difficult without such knowledge.

This book presents some of the important aspects of twentieth century physics in a form that should be accessible even to those with little previous knowledge of physics. It concentrates on the tremendous advances that have been made in the area that most physicists would agree is the central theme of this physics—the study of atoms and of their subatomic constituents. This focus on microphysics (atomic and subatomic physics) means that of the major twentieth century physical theories, we will be concerned mainly with the quantum theory and with certain aspects of the special theory of relativity, but not with the general theory of relativity. The latter theory deals with gravitational phenomena, which are quite important in the physics of some stars and of the whole universe, but which have had no significant role until now in what we know of microphysics.

I have purposely declined to follow a specifically historical approach to the subject. Although the topics are presented in approximately the order that they have been explored by physicists, I have not tried to present the ideas as they appeared to their discoverers. Instead, the standpoint is always that of present knowledge rather than that of the time of the discovery. While the history of physics is an interesting subject in its own right, I do not believe that relating the history of twentieth century physics is the best way to explain what we now understand, or even to show how the physicist goes about solving his problems.

The path we shall follow is the study of the properties of smaller and smaller objects, or, and as we shall see this amounts to the same thing, to describe the results of using more and more energetic probes to analyze the structure of matter. Most of the emphasis is on the new theoretical ideas that are required to understand the phenomena encountered on each level of experiment. In some cases, these ideas were already available when they were needed, as for example when the creation of subatomic particles was discovered in 1933. In other cases, the ideas were developed subsequent to the experiments, as was the quantum theory of light. In still other cases, we are not yet sure of the correct explanations: for example, the fact that so many distinct kinds of subatomic particles exist. Because of this, the story of microphysics is still incomplete, and thus still excites the interest of many talented scientists.

A brief historical summary of the topics to be treated is in order here. While the focus of twentieth century physics has been on the problem of the structure of matter, this problem is in fact a very old one, first raised by several of the Ionian Greeks in the sixth century B.C. Some of these philosophers believed that all forms of matter were composed of a few simple kinds, and that these simple "elements" were among the types of matter already known from ordinary experience—water, earth, etc. An alternative view was expressed slightly later by another group of Greek philosophers, the atomists, especially by Democritus. They agreed that all the observable kinds of matter were made of a few simple kinds, but believed that these primitive constituents of matter were quite different from the matter perceived by our senses, consisting of tiny, indivisible objects called atoms which were in constant motion through empty space. This version of the atomic hypothesis, as expressed in the poem "On the Nature of Things" by the Roman writer Lucretius, persisted for two thousand years. It was, therefore, a part of the general background of philosophical speculation when interest in the structure of matter was revived in the seventeenth and eighteenth centuries.

Since then the solution to the problem of the structure of matter has gone through several important stages. The first was the amplification and clarification of the ancient atomic theory so that it provided a simple model of the composition of ordinary matter on the smallest possible scale. This model was essentially completed by the middle of the nineteenth century, and when supplemented by the application of the Newtonian laws of the motion of bodies to atoms, it successfully explained many properties of matter, such as the relationship between the pressure and temperature of a gas. But in order to understand other properties of matter, especially those connected with electricity and magnetism, it was necessary to go beyond this simple model of featureless atoms to a study of what atoms are made of.

This was accomplished in the first quarter of the twentieth century. In this second stage, it was recognized that atoms are not indivisible, but themselves consist of smaller, electrically charged components called electrons and nuclei. Soon after this modification of the atomic theory, it was found that the motions of atoms, electrons, and nuclei must be described by different laws than the familiar Newtonian laws used to describe the motion of large bodies such as ping-pong balls or planets. The necessary laws to do this were discovered in the 1920s, and expressed in the form of a general description of physical phenomena called quantum mechanics. By applying quantum mechanics to the extended atomic theory, it became possible to understand the remaining properties of ordinary matter in terms of simple properties of atoms and their constituents. I have touched briefly on a few examples of these explanations in this book.

In the third stage, which continues today, physicists have turned to a study of these subatomic constituents which display a wealth of complex phenomena. Most of these phenomena do not influence the behavior of bulk matter under ordinary Earthlike conditions, but they nevertheless are of interest in their own right. This study of the behavior of the subatomic components of matter, known as particle physics, has been the main feature of the physics of the last twenty-five years. This shift in the interest of physicists to the study of the properties of the subatomic particles is the most recent example of the process of abstraction that has happened again and again in the history of science. For example, early astronomers turned from a study of the supposed influence of stars and planets on Earthly happenings to a study of the motions of these bodies as a scientific problem in itself. However, it should be noted that many physicists, in such fields as solid-state physics, continue to work on the detailed explanation of properties of bulk matter through the atomic theory and quantum mechanics.

Since many of the objects studied in particle physics are not found in ordinary matter, it has been necessary to develop elaborate and expensive equipment to produce and detect them. The development and use of this equipment is an integral part of twentieth century physics, and is discussed briefly also.

Author's Note

In this book, very large and very small numbers are written using exponential notation. That is, such numbers are expressed as a number between 1 and 10, multiplied by a power of ten. For example, the number five thousand is written as 5×10^3. The symbol 10^3 means the number one with three zeros following it, i.e., one thousand. Similarly, the decimal number 0.0004 is written 4×10^{-4}, where the symbol 10^{-4} means the number one divided by 10^4, i.e., one divided by ten thousand.

When numbers expressed in this notation are multiplied, the powers of ten are added. For example, $10^3 \times 10^4$ is 10^7, which expresses the relation that one thousand multiplied by ten thousand equals ten million. In a similar way $10^5 \times 10^{-4}$ is 10^1, which states that one hundred thousand multiplied by one divided by ten thousand equals ten.

The common names for some powers of ten are given below.

10^0=one

10^1=ten

10^2=one hundred

10^3=one thousand

10^4=ten thousand

10^5=one hundred thousand

10^6=one million

10^9=one billion

10^{12}=one trillion

10^{-1}=one tenth

10^{-2}=one one-hundredth (0.01)

10^{-3}=one one-thousandth (0.001)

10^{-6}=one one-millionth (0.000001)

The terms and concepts introduced throughout the book are also listed in the glossary. Each of the glossary entries is italicized in its first occurrence in the text.

WHAT IS THE
WORLD MADE OF?

Atoms, Leptons, Quarks,
and Other Tantalizing Particles

I

Physics Before the Twentieth Century

The state of knowledge in physics before the twentieth century might best be characterized by saying that physicists had an accurate way of describing the behavior of familiar objects, but had only a vague understanding of the ultimate composition of those objects. This general description was called Newtonian mechanics, dating from 1667 and devised originally to treat the motion of the planets. As later modified in the eighteenth century, Newtonian mechanics also proved capable, with some extra assumptions, of explaining a great deal of the behavior of familiar objects on Earth.

The composition of bodies was generally believed to be given by some version of the atomic theory. According to this model, which in its earliest form dates from Leucippus and Democritus in 500 B.C., all matter is composed of a few kinds of very small bodies called *atoms,* which in solid objects are held together by unknown forces, and in gases are relatively separate and free to move about. While the evidence for the atomic theory was gradually strengthened through the nineteenth century by the work of both physicists and chemists, there were still serious scientists in 1900, such as the physicist Ernst Mach and the chemist Wilhelm Ostwald, who found the evidence unconvincing, and rejected the theory.

Various other phenomena that were studied by pretwentieth century physicists included electricity and magnetism, light, and, at the end of the nineteenth century, X rays and radioactivity. In some cases, such as James Clerk Maxwell's theory relating light to electricity and magnetism, physicists had been able to discover important relations between these phenomena and to summarize these relations in systems of mathematical equations. However, the connection between the other phenomena and the structure of matter remained mysterious.

The Atomic Theory

Before the twentieth century, the idea that matter is composed of atoms had two distinct aspects. On one side was the chemists' conception of atoms, based on the ways in which different substances combined with each other. In order to justify the observations of the chemists, it is essential that each element contain many copies of one type of atom, all of which have approximately the same mass and other properties, and that there be definite rules to determine how many atoms of one element could enter into chemical reactions with those of another element. In these reactions the number and type of each atom does not change. Instead, the substances are able to form temporary attachments, known as *molecules,* and any sample of a compound chemical substance, such as water, or salt, is made up of many copies of the same molecule, consisting of a union of two or more atoms. The actual size, mass, and other detailed properties of individual atoms were not essential for explaining the basic facts of chemistry. Indeed, most of the processes studied by chemists even to the present involve immense numbers of atoms, and so are insensitive to the properties of individual ones.

The physicists' view of atoms was somewhat more speculative than the chemists', but also was more detailed, in that it has more to say about the properties of the atoms. It was based on the ideas that the behavior of matter under some circumstances followed certain simple laws that could easily be understood if the matter was made up of a very large number of small *particles,* behaving in about the way that was expected by Newtonian mechanics. For example, a gas in a container exerts an equal force on all sides of an object immersed in it, so that there is no tendency for the body to move in any direction at least on the average. However, the gas inside does exert an unopposed force, called a pressure, on the sides of the container if there is no gas on the outside. This can be measured if we try to move the side of the container inward, compressing the gas. To do this requires that an additional force be exerted on the side being moved, which must increase as the gas is compressed more and more (Fig. 1). This behavior can be understood by imagining the gas to be composed of a swarm of atoms moving freely in all directions. An object immersed in the gas will collide with many atoms, but equal numbers will hit from all directions, and there will be no average force on it. However, a side of the container is hit only by atoms from within and so an unopposed force acts on it.

This explanation of pressure was first given in the eighteenth century. However, only in 1905 did Einstein point out that it had an interesting sidelight. While it is true that the many atoms hitting an object inside the gas will give no average motion to it, nevertheless the object will have many small motions in random directions, because the atoms do not all hit it at once (Fig. 2). This should lead to a kind of jittery motion of small objects suspended in a gas, which Einstein thought might be observable. Indeed, it had already been observed one hundred years earlier and is known as the Brownian motion.

Brownian motion can be seen by viewing through a microscope the motion of small objects suspended in a liquid or gas. The size of the objects for which this can best be seen is about 10^{-4} centimeter (cm). Even such small objects are 10^4 times larger than the atoms that are colliding with them to make them move. The effect is about the same proportionally as the effect on a man's motion when he collides with a swarm of mosquitoes. What is observed is a slow drift of the suspended object through the liquid as a result of many random motions in different directions. For a sphere with a radius of 10^{-4} cm, the amount of drift under typical circumstances is about 10^{-2} cm in an hour, or a thousand times slower than a snail's pace.

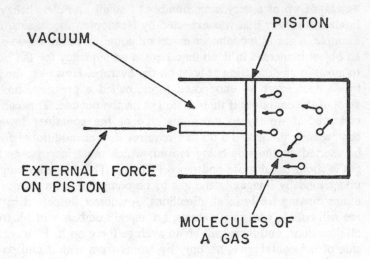

VACUUM

PISTON

EXTERNAL FORCE
ON PISTON

MOLECULES OF
A GAS

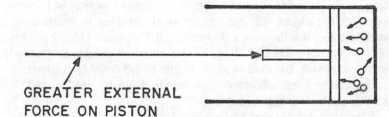

GREATER EXTERNAL
FORCE ON PISTON

FIGURE 1. Explanation of gas pressure by the atomic theory. The wall of a container, here represented by a piston, has a gas on one side, and no gas on the other. The molecules or atoms in the gas collide with the wall, exerting a force on it, which is not matched by collisions from the other side. This is observed as a pressure on the wall. When the gas is compressed, the molecules make more collisions in the same time, as they have less distance to travel. Therefore, the force they exert increases, and a greater outside force must be exerted to resist this force—hence the pressure is observed to increase.

MOVING MOLECULES
IN THE FLUID

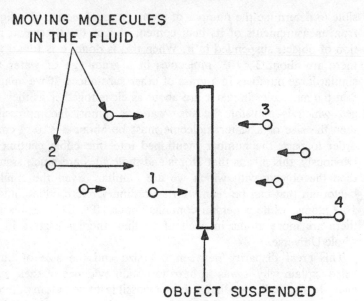

OBJECT SUSPENDED
IN A FLUID

FIGURE 2. Brownian motion. An object suspended in a fluid is constantly struck by the molecules of the fluid. Each time it is hit, it moves in the direction the molecule was going. The collisions occur at random times depending on the velocity and position of various molecules. In the diagram, the velocity of the molecule is represented by an arrow: the larger the arrow, the faster the motion. The suspended object will first be hit by molecule 1, then by 2, then by 3, then by 4. Hence it will first be pushed to the right-hand side, then to the left-hand side. This random motion is what is observed as Brownian motion.

The amount of motion is greater when the suspended object is smaller, and also when the temperature of the liquid is higher. Both of these would be expected from collisions with the atoms or molecules of the liquid. On the basis of measurements that have been made of the Brownian motion, and of Einstein's analysis of it, it is possible to infer the average energy of each atom, since the effect of each collision on the suspended body depends on the energy of the atoms colliding with it. Since the heat content of the liquid is essentially the total energy of all the atoms, it is then possible to determine the number of atoms in any amount of a liquid from measurements of its heat content and of the Brownian motion of objects suspended in it. When this is done, it is found that there are about 3×10^{22} molecules in a gram (g) of water and similar large numbers in a gram of other substances. If we imagine that the molecules in water are about as close together as they can get, which is plausible because water is almost incompressible, then the size of a water molecule must be about 3×10^{-8} cm in order to pack the number mentioned into one cubic centimeter. Obviously, this means that atoms and molecules are much smaller than the objects with which we are familiar. Even the smallest bacterium that can be seen with an ordinary microscope contains 10^{12} atoms, while a person contains about 10^{28}. This means that there are more atoms in each of us than there are stars in the whole Universe.

This great disparity between our size and the size of atoms helps explain why it was so hard to obtain evidence of their existence. It also indicates that it is not possible to see atoms directly by the use of light magnifying devices such as an optical microscope. The problem has to do with the structure of light itself. We will discuss later how each color of light has a length associated with it called its *wavelength*. These lengths are typically about 10^{-4} cm for visible light. Information about objects much smaller than this wavelength cannot be obtained no matter how the light is amplified or otherwise treated. This is because for such a small object the properties of a light beam will be approximately constant over the size of the object, whereas the formation of an image requires that different parts of the object should be illuminated by light of varying properties. Only by using X rays, which are related to light but with much smaller wavelengths, have physicists been able to obtain direct information about the way atoms and molecules "look," and even this information cannot be complete, for reasons we shall discuss later. A similar principle is involved in the use, by bats, of sounds of small wavelengths to guide their flight.

Because there are so many atoms in even a small sample of matter, the behavior and properties of matter depend mainly on average properties of the atoms. So we have no guarantee that all of the atoms in even a chemically pure sample of matter are identical. One of the accomplishments of twentieth century physics has been to clarify the extent to which atoms of the "same" substance are identical, and to what extent they may differ from one another. For example, it has been found that the atoms of what is chemically the same substance, hydrogen, occur in nature with two different masses, one of which is two times greater than the other. It has also been possible to understand how different arrangements of the same atoms can lead to different physical forms of the same chemical substance, such as the two very distinct forms of carbon, diamond, and graphite.

The recognition that atoms have a size, rather than being points, also comes from other evidence. When a gas is compressed, its density initially increases proportionally to the pressure applied. However, at high pressures, it gets harder and harder to increase the density, no matter how much pressure is applied. A simple way to understand this is that the atoms, or molecules, composing the gas have a finite size, and that when the density reaches a value such that the atoms are touching each other, a further increase in pressure cannot increase the density of the gas any further. Of course, such a statement can only be approximately true, as it is unlikely that the atoms are perfectly rigid. Indeed, we now know that at much higher pressures than are easily reached on Earth, the atoms begin to break apart into their constituents, and much higher densities can be reached. This is the situation inside many stars. From measurements of the density at which further increases in pressure on a gas produce no further increase in density, it is possible to infer the size of the atoms, and this again works out to be about 10^{-8} cm, the same value as inferred from the Brownian motion.

The finite size of atoms raises a very important question not previously apparent to believers in atomic theories. Our past experience has always been that anything whose size was not zero was made of still smaller things. It is therefore plausible to ask whether this is also the case for atoms. In addition, whether or not there are such subatoms, it is necessary to ask what the structure of atoms is. For example what is the distribution of the mass of an atom over its size? If atoms have parts, do these different parts move with respect to each other, and what happens to them when atoms combine chemically into molecules? The physics of the twentieth century has been devoted largely to questions of this type, and to a great extent we have been successful at answering them. In the process of doing so, we have found not only that the structure of matter is immensely complex on a subatomic scale, but also that the very physical theories we use to describe the atoms and their constituents are quite different from the Newtonian mechanics mentioned above. In the following chapters, some details about the composition of atoms will be given, and then the newer theories used to describe atoms and the components of atoms will be introduced.

Newtonian Physics

These newer theories have grown out of, and make use of, the concepts of pretwentieth century physics. As in most areas of science, an appreciation and understanding of newer concepts is made easier by some familiarity with what has gone before. Therefore, we shall set the stage for our survey of the accomplishments of twentieth century physics by discussing briefly some of the concepts of Newtonian physics that remain relevant to later discoveries although in an altered form.

The paradigm case of a physical system to which Newtonian mechanics was applied is a collection of small, structureless bodies, often called particles. These particles are imagined to move through definite paths in space, similar to the way the planets move in the solar system, or a thrown ball moves through the air. The motion of a particle at any time can be described by its velocity, which includes both its speed, and the direction of its motion. If there are several particles in a region, they can influence one anothers' motion, either through collisions in which they touch briefly, or through the action of forces such as gravity, by which the presence of one particle affects the motion of another at a distance from the first. According to what is known as Newton's first law of motion a particle that is very far away from other particles, so that it is essentially uninfluenced by them, would move in a straight line at a constant speed. Any deviation from such uniform motion is ascribed in Newtonian mechanics to the effect of other objects, and is said to be due to a force acting on the body that is moving differently. So for example, the fact that the Earth moves in an orbit around the Sun, rather than in a straight line, is ascribed to the force of gravity exerted by the Sun on the Earth. Motion that is not in a straight line at constant speed is known as accelerated motion, and the acceleration is simply the rate at which the velocity changes. Another way of stating Newton's first law of motion is that when there is no force acting on a body, it is not accelerated. Newtonian mechanics by itself did not attempt to explain what forces might exist in nature, but rather described how motion occurred when the force was known. Other branches of physics, such as the study of electricity and magnetism, attempted to determine what these forces actually were.

An important property of all types of forces considered in Newtonian mechanics is that the force exerted by one object on a second is equal in strength and opposite in direction to the force exerted by the second object on the first. That is, the influence of objects upon each other is mutual, not one sided. This has as a consequence one of the important aspects of Newtonian mechanics which has been preserved in twentieth century physics, the law of *conservation of momentum.*

Momentum is a quality with both a direction and an amount, associated with any moving particle. The direction of the momentum is the direction of motion. The amount of momentum is proportional to the speed of the particle, but different particles moving at the same speed may have different amounts of momentum. When this happens, the particles are said to have different masses. In Newtonian physics the momentum of any particle is therefore proportional to both its mass and to its velocity. Momentum, and also mass, are additive, in that the total momentum of two or more bodies is obtained by adding the individual momenta. However, since momentum like velocity, is a quantity with an associated direction, this addition is more complicated than it sounds. If the bodies are moving in the same direction, the total momentum is also in that direction, and its amount is the sum of the amounts. If the bodies move in different directions, it is necessary to analyze their motion in terms of motions along three fixed directions, and then add the components of momentum along each of these directions (Fig. 3).

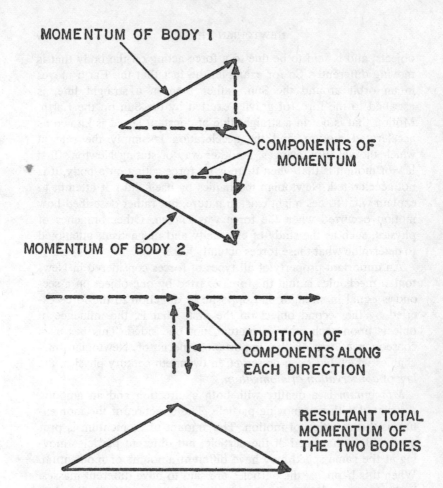

MOMENTUM OF BODY 1

COMPONENTS OF
MOMENTUM

MOMENTUM OF BODY 2

ADDITION OF
COMPONENTS ALONG
EACH DIRECTION

RESULTANT TOTAL
MOMENTUM OF
THE TWO BODIES

FIGURE 3. Components of momentum and their addition. The momentum of an object is represented by an arrow, pointing in the direction of motion, whose length is proportional to the amount of momentum. The momentum can be resolved into components along any three directions. In the diagram, a simple case is considered of an object whose motion is in the plane of the paper, so that only two directions are necessary, which are taken as the horizontal and vertical directions. When the arrows representing the components are placed head to toe and recombined, they give the arrow of the original momentum. To obtain the total momentum of two bodies, the components along each direction are added, and the sums combined by the head-to-toe method. This gives the same result as if the two momenta were added directly by that method. Note that the components may have opposite signs for two bodies, and so may cancel when added.

One may ask whether the total momentum of several bodies is of importance, if the bodies are not somehow part of the same object. By contrast, the total velocity of two bodies, say two cars traveling in opposite directions, has little significance. The answer is that for any collection of objects that are far from all other objects, and so uninfluenced by the other objects, the total momentum remains constant, no matter how the objects are moving. This is what is known as the law of conservation of momentum. If the collection consists of a single isolated particle the law says simply that its momentum is constant. Since we already know that its velocity is constant, this implies that its mass is constant as well. A single object that is not a particle, but has some internal structure, may not have a constant mass. For instance, it may break up into pieces, like a bomb, or continuously emit small parts of itself, like a rocket. In this case, the total momentum of all the parts, when added together, will equal that of the original object.

If two or more objects are involved, the momentum of each one may change because of the action of the other, but these changes compensate each other, and the total momentum remains fixed. This is the case for two particles held together by the gravitational force they exert on one another, or for two pool balls colliding on a frictionless table. Very accurate measurements have been made of the momentum before and after collisions of the very small particles that compose atoms, and it has been found that the law of conservation of momentum is satisfied in these collisions. Therefore, the law of conservation of momentum has been incorporated into the theories of twentieth century physics. In doing so, a minor change has been included. Momentum, although it still always has the same direction as the velocity, is not always proportional to the magnitude of the velocity, the speed. Instead, the amount of momentum depends on the speed in a more complicated way. This change, required by Einstein's special theory of relativity, is only important for bodies whose speeds approach the speed of light, 3×10^5 kilometer per second (km/sec). While this modification is not relevant for ordinary objects, or even for many atomic processes, it is crucial for many subatomic processes, where such speeds are common.

Another property of the forces considered in Newtonian mechanics is associated with an even more famous law, the law of *conservation of energy*. The energy of a Newtonian particle is somewhat more complicated than its momentum. One aspect, the *kinetic energy,* depends only on the mass and speed of the particle, but not on its direction of motion. This kinetic energy is proportional to the mass, and to the square of the speed, so that it quadruples when the speed doubles, etc. Unlike the momentum, whose component along some line can be either positive or negative, depending on the direction of motion, kinetic energy is never negative. The total kinetic energy of a system of particles is the sum of the kinetic energies of each particle.

If the only forces between the particles occur during collisions, and if the particles are truly structureless, so that they do not break apart or heat up during collisions, then this total kinetic energy is constant, in the sense that it has the same value before and after each collision, even though the kinetic energy of each particle may change. This is the simplest form of the law of conservation of energy. If, however, the particles exert forces such as gravity on one another even when they are apart, then it is found that the total kinetic energy is not constant. Instead, the total kinetic energy will vary with the distances between all the particles. In this case, what remains constant is the sum of the total kinetic energy and a quantity that depends on the mutual distances of all the particles, called the *potential energy.* This is because the forces that the particles exert upon each other result in changes in their kinetic energies, and the way that these changes in the total kinetic energy occur can be summarized by the concept of potential energy.

Perhaps the most familiar example of potential energy is that involving the gravitational force of the Earth. For some purposes we may consider the Earth, and an object near it, such as a falling apple, as two Newtonian particles. Experience shows that the speed, and hence the kinetic energy, of the apple increases as it falls. The change in kinetic energy of the Earth is so small it may be neglected. Therefore, the total kinetic energy of the two increases as the apple falls. This means that the potential energy must decrease. Experiments show that for an object falling in vacuum, the increase in kinetic energy depends only on the initial and final heights of the object, so that the potential energy also depends on this height. This can be checked by throwing the apple upward or downward at various initial speeds. Although the actual speed at any height above the Earth will depend on the initial speed, the change in kinetic energy between two heights is found to be the same, no matter what the initial speed, provided that the object is moving slowly enough that air resistance is unimportant. The potential energy in general depends on the positions of all the objects that are influencing each other, and is related in a simple way to the forces they exert on one another. The law of conservation of energy for systems of particles implies that while the total kinetic energy of the particles may change, and their potential energy may change also, these changes will balance each other. Alternatively, it can be taken to mean simply that changes in the total kinetic energy are uniquely specified by changes in the distances of the particles from each other. These two ways of phrasing the law really say the same thing, but the first is preferred by physicists, because it emphasizes that something is constant.

The General Law of Conservation of Energy

A still more general form of the energy conservation law comes when we consider objects more complex than particles, those with an inner structure that may change with time. For example, in collisions of actual bodies like pool balls, the bodies may become dented, warmed, or break apart. Further, when bodies are near each other, they may combine chemically, become physically attached, or emit light. One of the most remarkable discoveries of pretwentieth century physics was that even in complicated processes of this kind, it is possible to find an extension of the energy conservation law that remains true.

For example, when an electric motor is operated, a definite amount of electric current, at some voltage, is introduced into the windings of the motor, causing the wheels to spin at a certain rate. The spinning wheels have a kinetic energy, because each small bit of matter composing the wheel is in motion. Thus something about the electric current has been converted into kinetic energy. In a very efficient motor, meaning one in which little heat is produced, a definite amount of electric power, that is, voltage times current, acting for a definite time, will always produce the same amount of kinetic energy in the motor. It is therefore natural to associate this quantity—power multiplied by the time it acts— with the energy of the electric current, and to conclude that it is this electrical energy, when added to the kinetic energy of the wheel, that is conserved. This identification is strengthened by the fact that in an electric generator, which is essentially a motor run backward so as to produce an electric current, the amount of electrical energy produced, defined in the same way, will again be equal to the change of kinetic energy of the wheel in the motor. In actual motors and generators, these relations are not exactly correct, because heat is also produced. But the heat generated is also related to the kinetic energy changes, and to the electric energy. Because of this, and other observations, it is plausible to associate a definite amount of energy with a certain amount of heat, and to extend the conservation of energy to include this heat as well.

Another circumstance to which the energy conservation law can be extended is the production of heat, or of motion, through chemical changes. When some gasoline is burned in a car engine, a chain of events takes place, ending with the car increasing its speed, and some carbon dioxide, water, and other wastes being exhausted. The chemical change of gasoline and oxygen into these wastes results in the kinetic energy of motion of the car. So if energy is to be conserved, there must be a different energy content in the original gasoline and oxygen than in the waste products. This amount of chemical energy, as it is called, can be determined by measuring the amount of heat produced in the burning of the gasoline, since we already have seen that heat has a definite energy. It is found that every chemical change occurring under definite conditions of pressure, etc., leads to a specific amount of heat produced, or in some cases absorbed, when it occurs. The amount of energy involved is proportional to the amounts of the substances involved in the chemical change. However, this energy must be thought of as associated with the whole chemical change, rather than with each substance, because different chemical reactions involving the same substances can produce different amounts of heat.

In the form that I have described it, which is about how it was understood in 1900, it would appear as if the concept of energy was not a very definite one, but rather could be indefinitely extended to include as many forms as necessary to make the conservation of energy be true. One of the successes of atomic and subatomic physics has been to reinterpret the meaning of energy so that the many different forms known before can be identified as essentially just the kinetic or potential energy of the atomic and subatomic constituents of matter. For example, even when an object appears at rest, the atoms composing it are moving in a random fashion within the body, and the kinetic energy associated with this motion is the heat energy of the body. Similarly, the chemical energy of a reaction can be understood as the change in the kinetic and potential energy of the atoms composing the substances involved in the reaction, between their initial configuration, say in gasoline and oxygen, and their final configuration, in carbon dioxide and water. This reinterpretation of the meaning of energy has been a great help in simplifying the physicists' picture of the composition of matter. Furthermore, the energy conservation law, especially in the simple form involving the potential and kinetic energy of particles, has been adopted, again with a minor change that is required by Einstein's special relativity theory, into the theories invented in the twentieth century to describe atoms and their constituents.

This is not the case for another law believed to be true before the twentieth century, the law of conservation of mass. This law states that in any process involving bits of matter, either particles or composite bodies, the total mass of all the bodies remains constant. For example, if two bodies collide and stick together, the total mass of the collision product would be the sum of the masses of the bodies that collided. This law is approximately true, in that for most processes that occur among bodies of ordinary size, and even among atoms, any change in mass is too small to observe. However, in certain subatomic processes that we shall describe later, the mass can change substantially. This possibility is related to a consequence of one of the new theories introduced in twentieth century physics, the special theory of relativity. We shall see that in this theory, a body at rest has an energy content proportional to its mass, and that it is possible for the total mass of bodies involved in some reaction to change, with the difference being made up by a change in the total kinetic energy of the bodies. So strictly speaking, the two laws of conservation of energy and of mass are replaced by a single law of conservation of energy, in which part of the energy of a body is proportional to the mass. In most cases, the separate part of the energy involving the mass, known as the *rest energy,* does not change appreciably, so that there is effectively a conservation law for mass as well as of the new form of energy. But in the subatomic processes mentioned this is not the case, and the total energy is conserved.

One more aspect of pretwentieth century physics that has played an important role in later developments has been the properties of electric and magnetic forces. After a twenty-five-hundred-year-long investigation of various aspects of electricity and magnetism, a grand synthesis emerged in the late nineteenth century in the work of James Clerk Maxwell and Hendrik Lorentz. Maxwell's main contribution was a set of mathematical equations that describe how electric and magnetic forces influence one another as they vary in space and time. Lorentz helped to clarify what has been suspected before about the ultimate origin of these forces. According to his ideas, there are certain particles endowed with an additional property called electric charge. When these bodies are at rest, they exert forces on one another, even when the bodies are not touching, which we identify as electric forces. When the particles move, they exert other forces, in addition which we identify as magnetic forces. The electric charge occurs in two types, positive or negative, and these two types produce opposite electrical forces on another charge, so that if a body contains equal numbers of the two kinds, it is neutral and exerts little electrical force. However, even in a neutral body, if the two types of charged particles inside the body are moving differently, they can produce a magnetic force, and in this way a body that is overall at rest, such as a bar magnet, can still show magnetic effects (Fig. 4).

In this theory, the total amount of electric charge must remain constant, whatever happens to the particles that carry it. This is called the law of conservation of charge. When an amount of positive charge is produced, say by friction, an equal amount of negative charge must be produced at the same time. This suggests that the charges are both present in a combined form in matter before

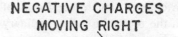

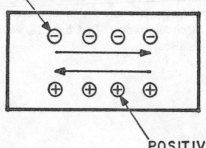

POSITIVE CHARGES
MOVING LEFT

FIGURE 4. Current flowing in a neutral object. If a body contains equal numbers of positive and negative charges, the total charge vanishes, and the electrical force it exerts is greatly reduced. However, if the positive and negative charges move in opposite directions, there is a net flow of electric current, even though the body as a whole remains at rest. This current exerts a magnetic force on another such current.

the separation, a result that was soon verified, as we shall see in Chap. II. Once the two kinds of charge are separated, they can easily move from place to place. This moving charge is what we call an electric current, and the amount of current flowing is proportional to the amount of moving charge multiplied by the speed at which it moves. The electrical energy of the current, which we have seen previously can be transformed into kinetic energy or heat energy, is in this way of looking at things simply the potential energy of the moving charges, which in the present case depends not only on the distances between the charges but also on their velocities, because of the magnetic forces. Actually, the energy of a current is both magnetic and electric energy, but the latter phrase is too common to change. According to the equations written by Maxwell, the two kinds of force are really aspects of one phenomenon, which is now called electromagnetism, so that the distinction is not too important.

With the work of Maxwell and Lorentz, the understanding of electromagnetism was fairly complete, except for the crucial question of the relation between electric charges, and other forms of matter. The key discovery came with the recognition that ordinary matter consists of electric charges bound tightly together into atoms by electrical forces, so that the matter appears electrically neutral under most circumstances. In Chap. II, we shall examine the evidence for this discovery.

II

What Atoms Are Made of

It is strange, but perhaps not too surprising, that a general agreement among physicists about the existence of atoms occurred at about the time that evidence was being accumulated demonstrating that atoms are not the structureless particles imagined by Democritus, but instead have complicated internal structures. This is not surprising because it is common in science that belief in some entity or phenomenon is the result of a detailed study of its properties and its relations to other phenomena, rather than the result of some critical experiment that demonstrates its existence directly. The advances that have been made in our understanding of atoms in the twentieth century have come more from the analysis of the constituents of atoms, and how these affect atomic properties, than from inferences about atoms based on observations of bulk matter of the type described in Chap. I. Indeed, some properties of bulk matter, such as the electrical properties of metals, can only be understood by recognizing that one atomic constituent, the *electron,* can exist in bulk matter without being associated with individual atoms. It is, however, difficult to learn about electrons from this. More direct evidence about what atoms are made of first led to a knowledge of electrons and other atomic constituents.

There are at least two general circumstances in which atoms

break apart and their constituents become sufficiently separated that they can be studied independently of the atoms. For some atoms, this can happen spontaneously, without any outside stimulus to the atom. This involves a phenomenon known as *radioactivity* discovered in 1896 by Antoine Henri Becquerel, and intensively studied thereafter. Atoms can also be induced to break up into their constituents by causing them to collide with one another at high speeds. In these circumstances, just as in collisions between familiar objects, pieces of the atom emerge which are apparently more resistant to destruction than the whole atom. Furthermore, in the collisions that were studied in the early twentieth century, only a few different kinds of subatomic pieces ever emerged, rather than a different set of pieces from each collision, as might occur if a glass is thrown on the floor. It therefore is natural to assume that these few kinds of pieces were present in the atoms before the collision, and are being liberated, rather than created. We shall see however that this seemingly reasonable assumption cannot be extrapolated to all circumstances, particularly those more violent collisions being studied by contemporary physicists.

In order that atoms should break up in a collision, the kinetic energies of the atoms must be somewhat higher than the energies atoms have under ordinary Earthlike circumstances. If this were not the case, whole atoms would rarely be found on Earth, since they are always colliding with one another. This is actually what happens inside stars, where the atoms have much higher energies because the temperatures are higher. Furthermore, when the atoms collide at high velocities, some of this velocity is transferred to the subatomic parts that emerge from the collision, allowing them to escape more easily from the matter containing the atoms, and so be detected. If the collision products are moving too slowly, they are likely to be recaptured by other atoms before they can be detected.

In order to give atoms energies and velocities substantially higher than normal, the atoms must be accelerated. This is most easily done if the atoms are given an electric charge. This can, for example, be done by passing an electric current through a gas. Since atoms normally have no electric charge, giving them one implies that something with an electric charge is added to or subtracted from them. But this in itself tells us little about what these charges are, or whether they are constituents of normal atoms at all.

An electrically charged atom, or *ion,* when acted on by an electric force, will be accelerated in the direction of the force. If the ion is moving slowly at the beginning, and the force acts on it for some time, the ion will eventually be moving mostly in the direction of the force, as its velocity will be mainly the result of its acceleration by the force. Another way of saying this is that the energy of motion (kinetic energy) of the ion will mostly have come from the work done on it by the accelerating electrical force. If this force is approximately constant in strength, the work done on the charge, which is just the change in the kinetic energy of the charge, will be proportional to the distance that it moves while the force is acting on it. The work will also be proportional to the strength of the force. The product of force with distance depends on what potential difference the charge moves through, which can be measured in volts (V). In many cases the change in the kinetic energy of the charge can be thought of as balanced by a compensating change in its potential energy, with respect to the other electric charges that are generating the electrical force that is acting on it. The change in kinetic energy of the charge also depends on the amount of electric charge it has. We will see that in almost all cases, this amount is one or two times a universal unit of charge, that associated with a subatomic constituent called the electron. The amount of work done on the unit amount of charge moving through a potential difference of one volt is called one *electron volt,* or 1 eV. By the conservation of energy, this also corresponds to the kinetic energy obtained by such a charge moving through that potential difference. It turns out that this is a convenient unit of energy to describe atomic collisions and atomic processes, because atoms with energies of a few electron volts can often be broken up when they collide. It should not be thought that whenever an atom has 1 eV of kinetic energy, it is the result of being accelerated by such an electric force. As mentioned in Chap. I the whole point of the energy concept is that different forms of energy are interconvertible and only the total energy remains constant. To say that an object has 1 eV of kinetic energy therefore means only that if it is given a charge of one unit, and allowed to move backward through a potential difference of one volt, its energy would be decreased to zero. It may have acquired this energy in various ways, such as collisions, other types of force, etc.

A stream of moving ions is similar to an electric current, although it differs somewhat from the electric currents in wires, which involve the motion of electrons. One place in which the effects of such moving ions can be studied is in a device known as a discharge tube. This consists essentially of a long tube, containing a gas at low pressure, with metal pieces, called electrodes, at each end connected to a source of electricity. One electrode, the cathode, is kept at a lower voltage than the other, called the anode, by an external source of electricity. The familiar neon light bulb is a common example of a discharge tube. Many remarkable phenomena are observed as the voltage difference, type of gas, and gas pressure are varied, and these phenomena have led to, among other things, the discovery of X rays and of electrons.

Two of the important phenomena observed in a discharge tube are the cathode rays and the positive rays. These are observed as streams of light which extend from one end of the tube to the other. However, the ray itself is not a light ray, but instead produces the light by colliding with atoms in the gas. The cathode rays give off blue light and travel in straight lines away from the cathode. The positive rays travel toward the cathode. These behaviors are what would be expected if the cathode rays consist of negatively charged particles, and the positive rays of positively charged particles, as the electric forces act oppositely on the two charges. A more direct proof of this comes by introducing additional known electric and magnetic forces to act on the rays from outside the tube, and noticing that the rays are accelerated by these forces in the way that charged particles should be.

Electrons and Positive Ions

The positive rays are the streams of ionized atoms we have mentioned above. On the other hand, the cathode rays are not atoms that are negatively charged. There are several reasons for believing this. The best evidence comes from a measurement of the deflection of the cathode rays by electric and magnetic forces. The deflections of a charge by such forces depends on the amount of charge, on the mass of the charge, and on its velocity. By measuring the deflections, the velocity and the ratio of charge to mass can be determined, but not the charge and mass separately. While the velocity is found to vary with the conditions of the discharge, the ratio of charge to mass of the cathode-ray particles is always the same whatever the gas, strongly suggesting that a unique type of charged object is involved in them. These objects are called electrons, and their properties are found to be the same whatever their origin. It is plausible that the electrons are ejected from the cathode by collisions of the rapidly moving positive ions with the atoms in the cathode. The electrons then are repelled from the cathode by the electric forces acting on them, and eventually collide with neutral atoms in the gas of the tube. These collisions tend to ionize the neutral atoms, leaving them with a positive charge, after which they are attracted to the cathode, and eventually eject more electrons. In this way the cathode rays and positive rays maintain each other and the other aspects of the discharge. The exact mechanisms by which these collisions give the effects they do are complicated, but it is plausible that when charged bodies are close to another, the strong electrical forces they exert on one another would tend to produce rapid motions of each body.

The same measurement procedure can be used for the positive rays, and for them it is found that the ratio of charge to mass varies with the nature of the gas contained in the tube; hence it is more likely that different objects are involved in each case. Furthermore, the charge-to-mass ratio is much greater for the cathode-ray particles than for any of the positive rays. The charge of the cathode-ray particles and of the atoms can be measured separately, and are found to be equal in magnitude, within the accuracy of measurement, leading to the conclusion that the electrons are much less massive than atoms. The fact that electrons and positive ions have charges that are equal in magnitude, or at least simply related, can be seen without direct measurement. If this were not the case, it would not be possible for neutral atoms to be composed of electrons and positive ions, as the charges would not add to zero. But since the cathode rays and positive rays originate in matter that is originally neutral, it is plausible that they have equal but opposite charges. The point of this argument is that there is no necessary relation between the masses of electrons and ions, because the masses of neutral atoms are all positive, whereas the vanishing charge of these atoms makes it necessary that their constituents have simply related charges. The same reasoning suggests that all electrons have charges that are either the same, or simple multiples of one another. Actually, the charges of all electrons are equal, while positive ions may have charges equal in magnitude to one or more electron charges, depending on how many electrons have been removed from a neutral atom to make the ion. The numerical value of the charge of an electron is symbolized by $-e$, the minus sign indicating that it is negative.

The variation of the charge-to-mass ratio for different ions is then used to determine the mass ratios of atoms, since very little of the mass can come from the much lighter electrons, and, once the charge of any one ion is known, to determine the absolute mass of atoms. The values obtained for atomic masses in this way agree with the values obtained more indirectly as discussed in Chap. I. It is found that the mass of the lightest atom, hydrogen, is approximately 1.6×10^{-24} grams (g), and that of the electron is about 10^{-27} g. Thus a gram of hydrogen contains a huge number of atoms, about as many atoms as the number of stars in all the galaxies in the known universe. The masses of atoms are often instead expressed as ratios, with the mass of a carbon atom being arbitrarily taken as twelve units. The numbers obtained in this way are known as the atomic weights of the atoms.

The idea that electrons are constituents of normal matter, and hence of atoms, is reinforced by other observations. One is the so-called photoelectric effect, which involves the observation that when violet or ultraviolet light strikes a metal surface, the metal tends to become positively charged, suggesting that negative charges are ejected by the light, since the light rays themselves carry no charge. This is known because light rays are unaffected by electric forces. The ratio of charge to mass for these negative charges can be measured, and is found to be the same as that of the electrons in the cathode rays, suggesting that electrons are being ejected from the metal by the light.

Still another method of liberating electrons from matter, particularly certain metals, is by heating the matter. At temperatures of several thousand degrees, a small fraction of the electrons are given enough energy, through collisions with the rapidly moving atoms in the metal, to be ejected. The process by which this happens is very similar to the evaporation of atoms from a solid body that is heated. This emission may be observed by noticing a flow of electric current from the heated metal. The effect is known as thermionic emission, and has been used as a source of electrons in the vacuum tubes that were used until recently in television sets and elsewhere.

The picture of atoms emerging from these observations is, then, that atoms are composed, at least in part, of two kinds of bodies carrying positive and negative electric charges, with the negative charges being always the same and the positive charges varying in mass and perhaps in charge from atom to atom. The next questions that occur to one are how many charges of each type are contained in an atom, and what the arrangement of charges is within the atom. Before turning to these questions, let us consider another source of information about the constituents of atoms, the spontaneous breakup of certain atoms known as radioactivity.

Radioactivity

Radioactivity was first observed through the emission of some kind of radiation from certain materials such as uranium; this radiation, like ordinary light, could affect a photographic plate, but unlike ordinary light could pass through dark paper, and through thin layers of metal. These radiations produce ions when traveling through gases, and also cause fluorescence, an emission of visible light, when they strike certain materials. Both of these properties are used to detect the radiations. Among those materials occurring naturally on Earth, only a few of the elements, mostly those with high atomic weights are found to be radioactive. However, by means of the artificially induced transformations of elements that we have learned to accomplish in this century, it is found that most other elements have forms that are radioactive. The reason that these are not found in nature is that they emit their radiations rapidly and quickly change into other nonradioactive forms. As a result, any amounts of these elements that were present at the formation of the Earth have disappeared long ago. A few radioactive elements with large atomic weight emit their radiations more slowly, and some of the original amounts of these remain on Earth. If the Earth were one hundred times older than it is, those too would be mostly gone, and it would have been difficult to discover radioactivity at all.

By experiments similar to those with cathode rays and positive rays, it can be shown that several types of radiation occur in radioactivity (Fig. 5). One type, called alpha rays, consists of positively charged objects, with the same mass as that of a helium atom, and a charge of twice that of a hydrogen ion. These alpha

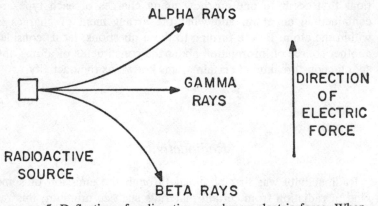

FIGURE 5. Deflection of radioactive rays by an electric force. When an electric force is made to act on the rays emitted by a radioactive material, three different effects are noted. The alpha rays are deflected slightly in one direction, indicating a positive charge. The beta rays are deflected by a larger amount in the opposite direction, indicating a negative charge. The gamma rays are undeflected, indicating that they are uncharged. Any one kind of atom will usually not emit all three radiations, so the source may be imagined to contain several types of radioactive atoms.

rays, or *alpha particles,* as the constituents of the radiation are called, penetrate only small amounts of matter before losing their energy of motion. Their identity to helium was shown by allowing many of them to come to rest in an evacuated tube, and showing that there was helium present in the tube afterward.

A second type of radiation, the beta rays, is found to consist of particles with a negative charge, equal to that of an electron, and approximately the mass of an electron. These beta rays are therefore streams of electrons. A measurement of their velocity indicated that the electrons in the beta rays were traveling at nearly, but slightly less than, the speed of light, much faster than the electrons in discharge tubes travel. The corresponding kinetic energies of the electrons were as high as several million electron volts (MeV), much higher than those of electrons produced through other means. Similar high values were obtained for the energies of the alpha particles, although their velocities were somewhat lower than those of the electrons, because their mass is much greater. The higher velocity of the beta rays allows them to penetrate through larger thicknesses of matter than the alpha rays. In both cases, there is a relation between the energy of the particle and its range, or the amount of a given type of matter it can penetrate. These relationships provide, once the relation has been determined for particles of known energy, a simple way to measure the energy of unknown examples of either radiation by measuring their range in matter.

When this is done, it is found that a specific radioactive element will emit alpha particles of a definite energy, or at most a few distinct values of the energy. On the other hand, the beta-ray electrons emitted by a particular element can have a continuous range of energies, up to some maximum value that is unique for each element. We shall see that this difference in behavior can be accounted for by the fact that the emission of beta-ray electrons is accompanied by emission of another particle, difficult to detect, known as the *neutrino.*

Another interesting property of the beta-ray electrons is that their mass, as it appears in the charge-to-mass ratio measured by deflection experiments, depends on their energy or velocity. The amount of dependence is small at low speeds, however, as the speed increases, the mass increases, and becomes indefinitely great

as the speed approaches that of light, about 3×10^5 km per second (km/sec). As a result, an electric field is less effective in deflecting a rapidly moving electron than a slowly moving one. These effects also occur for other rapidly moving particles, but it took longer to demonstrate this because the velocities of these particles, even when produced in radioactive emissions, are much less than the speed of light. A quantitative understanding of this variation of mass with velocity is given by the theory of relativity, which we shall discuss in Chap. V.

Finally, radioactive elements emit a third kind of radiation, called *gamma rays,* that is unaffected by electric and magnetic forces, and thus does not consist of charges at all. These gamma rays are similar to X rays in that they are a form of electromagnetic disturbance traveling at the velocity of light. They are of the same general nature as light rays, but have a much shorter wavelength, and so ordinarily do not behave much like waves. However, it is possible to show the wave properties of gamma rays and X rays by passing them through crystals, in which there is a regular arrangement of atoms. The gamma rays are reflected when they hit the crystals at very small angles, and the pattern of reflections depends on the wavelengths, which are found to be as small as 10^{-10} cm, or a million times smaller than that of visible light.

The gamma rays are able to eject electrons from atoms that they strike, in analogy to the photoelectrons ejected by ultraviolet light. In this way, gamma rays can ionize gases through which they pass, can affect photographic plates, and can be detected. However, the extent to which they do this is substantially less than for charged particles, and gamma rays penetrate through matter more easily than alpha rays or beta rays. We shall see later that gamma rays, like other electromagnetic disturbances, are made up of particlelike objects called *photons.* When a gamma ray is absorbed by matter, ejecting an electron, the basic process is an absorption of one of the photons by the atom. By measuring the energy of the ejected electron, it is possible to estimate the energy of the photon. For typical gamma rays, this energy is also about one MeV. Therefore, we see that all of the particles, charged or uncharged, emitted in normal radioactive transformations have energies around one MeV, or many times greater than the energies of the particles produced by thermionic emission or in dis-

charge tubes. This property has two consequences. One is that it suggests that the radioactive processes are quite different in nature than those processes involved in ordinary changes of matter. The other is that these high-energy particles have been useful probes for the study of other properties of matter, and indeed were the best available until around 1930, when machines to accelerate particles to high energies were first developed.

We have seen that alpha particles are helium ions, so that alpha radioactivity involves the production of the element helium from other elements. This already indicates that something else than ordinary chemical changes are occurring, since those changes never involve transmutation of elements. The element left behind after radioactive *decay* is also sometimes chemically different from the original one. For example, radium, an element that was discovered through its radioactive decay, is a metal which emits alpha particles. As a result, it changes into radon, another element that is a chemically inert gas. Also, the atomic weights of radium and radon differ by four units, which is the atomic weight of the helium ion that is emitted. This suggests that the radium atom breaks up into two pieces: a radon atom, and an alpha particle. The lighter alpha particle emerges quickly while the radon atom remains in the sample of radium, until it can escape by ordinary diffusion, which takes much longer.

Similar changes of elements occur in beta radioactivity. The atomic weight of the decaying atom does not change appreciably in this case, since the mass of the emitted electron is so small. However, there are changes in the physical and chemical properties of the material, indicating that a new element has been produced. Finally, gamma-ray emission does not result in a change in the element involved, which is consistent with the gamma ray being uncharged and having no mass. We shall see that there is strong reason to believe that neither the beta-ray electrons nor the gamma-ray photons are present in the atom before the radioactive decay occurs, and that, instead, they are created at the time of decay. These types of radioactivity therefore give only indirect information about the structure of the atom.

The radioactive decay of any type of atom takes place at a rate specific to that atom. This is not as easy to determine as it might seem for several reasons. If we begin with a pure sample of some element that is radioactive, some of the atoms will decay to atoms of another element in any short time period. In many cases, these new atoms will also be radioactive, and themselves decay, again perhaps into radioactive atoms that decay further. If only the total radioactivity of the sample is measured, say by detecting the total number of alpha, beta, and gamma rays that emerge in some time period, that will give a complicated combination of the radioactive decays of the different atoms that have been produced in the series of decays that began with the original atom. If, as is often the case, the atoms produced later in the series, called daughters, decay more rapidly than the parent atom, then most of the radiation seen will come from them. In order to be sure that the decay of a specific atom is being studied, it is necessary to separate it chemically or otherwise from all its daughter atoms, and parent atoms, and make measurements on it during the time period before the daughter atoms have had a chance to build up again in number. This is not always easy to do, but it has been carried out.

A very important result of the analysis is that in a sample containing one specific type of atom the fraction of the atoms that decay in any time period is a constant, independent of the chemical or physical environment of the atom, such as whether the atom is alone or near other atoms in a chemical compound, of the temperature of the sample, etc. The constant can be expressed as the length of time it takes on the average for half of the atoms in a sample to undergo radioactive decay. This is known as the *half-life* of the element. Half-lives of different atoms are known to vary over many orders of magnitude, from 10^{-8} seconds (sec) or less to 10^{20} sec or more, and give important information about the processes occurring inside the atom during radioactive changes.

The time dependence of radioactive decays indicates that the decay of a radioactive atom is largely independent of the previous history of the atom, as well as its present surroundings. Suppose we start with one million atoms of an element whose half-life is one year. In the following year approximately 500,000 of those atoms will decay, leaving 500,000. In the next year, half of the 500,000 will decay, leaving 250,000, etc. The atoms left after one year have no record of having survived one year already, and decay at the same rate as fresh atoms do. This is the case both for atoms that have survived since the origin of the universe and for "new" atoms that can be synthesized by methods we shall discuss later. In either case, the fraction of the remaining atoms decaying within the next half-life is the same. In this respect, atoms do not behave like people, for whom the death rate increases with the amount of time elapsed since birth.

This behavior of atoms implies that each atom decays at a random time, unaffected by the decay of other atoms. The regularity involved in the very idea of a half-life is then a consequence of the presence of many atoms in the usual samples that are observed. If only two or three atoms were observed, they might all decay within one half-life or none might decay, although if we waited for several half-lives it is likely that all would decay in that period. This is analogous to the flipping of coins. A single coin gives heads or tails randomly. In several trials, it may give all heads, or all tails, or any mixture of them. But in a million trials, the coin, if well made, gives approximately equal numbers of each, and the next flip will give a result independent of the results of the previous ones. The decay of individual radioactive atoms is the most random process we know because it is unaffected by any environmental, or other, influences. Yet it might be thought that, as in the case of a coin, the precise time at which the atom will decay is decided by some process occurring within the atom, and that by sufficiently careful study, we might be able to determine in advance when that decay will occur.

Nothing we have learned about atoms and about radioactivity in the twentieth century has furthered the realization of this hope, and it has been generally abandoned. If there really were some condition of the atom that decided when it would decay, then we would expect that large samples of atoms of a specific type that had been made in different ways, or were in different environments, would show some variations in their behavior, either giving different half-lives, or perhaps having a completely different time dependence of the decay. Neither of these are observed to occur. Consequently, physicists have been driven to the surprising conclusion that the time of decay of an individual radioactive atom is not determined by anything about the atom, but is instead a truly random process. We shall see in Chap. IV that this conclusion is an essential part of the description of atoms and their constituents by *quantum mechanics*.

The evidence of radioactivity, and of electrical discharges through matter, indicates clearly that atoms are not the indivisible points Democritus imagined, but instead are complex structures, composed at least in part of electrically charged components. The recognition that atoms, and hence material bodies, are made up of electric charges has immediate consequences for the understanding of electricity and magnetism themselves. One consequence is that electricity is not something different from matter, as had been thought by many nineteenth century physicists. Instead, we can say that matter as we know it is itself a manifestation of electrically charged particles, neutral in bulk because it contains equal numbers of positive and negative charges. Indeed, it is possible to give a systematic account of the electrical and other properties of bulk matter on the basis of its composition in terms of charged particles. For example, electrification of matter by friction, the first known indication of electricity, is to be understood as the transfer of electrons from one piece of matter to another. Conduction of electricity in metals is a flow of electrons that are detached from atoms in the metal from one part of the metal to another. In these and other processes which had previously been thought of as a motion of one substance—electricity—through another substance—matter, the simple picture now available is that it is really a part of the same matter that is doing the moving. In this sense, the discovery that matter is made up of charges represents a reduction of the number of independent things contained in the world.

The magnetic properties of substances such as iron can also be understood in terms of this picture. It has been known since Hans Christian Oersted's and André Marie Ampère's work in 1821 that an electric current produces magnetic forces. If there are moving electric charges in atoms, these can act as microscopic currents and produce magnetic forces, even when there is no flow of charge from one atom to another. This was also suspected by Maxwell and others in the nineteenth century, but they had no evidence that there were really moving charges inside matter. A detailed quantitive explanation of the type of magnetism displayed by various substances can be given along these lines, so that the magnetism of matter is also removed as an independent element of nature.

The realization that electricity, magnetism, and ordinary matter are all manifestations of the same objects is but one in a long series of cases in which physicists have found unifying elements between phenomena previously thought distinct. It is this type of discovery that has encouraged physicists to believe that there may be some ultimate theory unifying all natural phenomena.

Physicists next set about to understand the composition of atoms in greater detail, using first those probes furnished by nature, and later constructing machines that could produce more efficient probes of the structure of atoms.

The Distribution of Charges in Atoms

The first questions that needed answers were about the number and distribution of the charges in atoms. Since the atoms are electrically neutral, the total positive charge must equal the total negative charge, but the number of individual charges of each type need not be the same. In order to detect the individual electrons within the atom separately from the positive charges, it is possible to use the fact that electrons are much lighter than the positive charges. If an equal force is exerted on both, the electrons will accelerate much more than the positive charges do. Such accelerating charges will emit electromagnetic radiation at a rate depending on the number of charges and on their acceleration. By measuring the rate of radiation for a known acceleration, it is possible to determine the number of electrons in a known mass of some element. Since the number of atoms in this mass can be found from the weight of an atom, this measurement then determines the number of electrons in an atom. Such measurements have been done especially in elements of low atomic weight, using electromagnetic forces of known strengths to produce the acceleration of the electrons. The result is that the number of electrons in an atom is approximately but not precisely proportional to the weight of the atom, and that this number varies from one for a hydrogen atom up through ninety-two for the heaviest naturally occurring atoms, those of uranium.

On the assumption that the only negative charges in atoms are electrons, which is perhaps justified on the basis that electrons are the only negative particle that have been observed emerging from ordinary matter, we can obtain an independent estimate at the number of electrons in an atom. This is done by measuring the total positive charge, taking that as equal to the total negative charge, and then equating that number to the charge on an electron times the number of electrons. The total positive charge in an atom was first measured by Ernest Rutherford and some of his co-workers using alpha particles as a probe. A beam of alpha particles was allowed to pass through thin sheets of material, and the number of alpha particles that were scattered, or deviated from their original path through some fixed angle, was measured by detecting the alpha particles with a fluorescent screen. The amount of scattering was found to depend on the material, and for a fixed angle, to increase with the atomic weight. Furthermore, it was found that the alpha particles were sometimes deflected through quite large angles, presumably by their collisions with atoms.

In order to understand these results, we must ask what structures within the atom are scattering the alpha particles. It is easy to see that the electrons cannot do much to scatter them, because of the huge difference in mass. It would be like trying to deflect a bowling ball by putting a ping-pong ball in its path. Similarly, if the positive charge was divided into a large number of pieces, each of which was relatively light, the alpha particle would never be deflected by very much in a collision. Large deflections in a collision, as when two billiard balls collide with one another, can occur if there is at least one region in the atom containing amounts of mass comparable to or greater than that of the alpha particle. There might be several such centers of positive charge, or nuclei, in an atom. However, in that case, a problem would exist about how to hold them near one another in spite of the electrical repulsion that would exist between them. The simplest, and correct, assumption is that only one such region of positive charge, called the nucleus, exists in each atom, and that most of the mass of the atom is concentrated in the nucleus.

From that assumption, and the assumption that electrical forces act between the nucleus and the alpha particle, it is possible to calculate by the methods of Newtonian mechanics the path of an alpha particle that passes near the nucleus. This path depends on how near the alpha particle gets to the nucleus, something which is not easily determined in advance. Therefore, it is assumed that the particle has an equal chance of passing the nucleus at any distance. This in turn implies various probabilities of paths for the alpha particles, and of angles of deflection. The deflection is greatest for alpha particles that pass very close to the nucleus (Fig. 6), because the force is greatest in that case. If the experiment is repeated for a large number of alpha particles, the number deflected or scattered through various angles should follow the calculated probability pattern. This is found to be the case. Furthermore, the probability of being scattered through some specific angle is proportional to the square of the nuclear charge. This is because the more charge there is on the nucleus, the greater is the force between nucleus and alpha particle, and the greater the average deflection that the particle will get. Hence the same measurements, done in different materials, should determine the nuclear charges of these materials. It is found that this charge agrees reasonably with the number of electrons, as determined by the method described above, confirming the idea that all of the positive charge is concentrated in one region. We shall discuss the size and other properties of the nucleus further in Chap. VI. However, it is already clear from the fact that alpha particles can pass through matter to some extent that the size of a nucleus is much smaller than the size of an atom. If this were not so, there would be no space through which the alpha particles could go, since we already know that in solid materials the average distance between the atoms is comparable to the size of the atoms. Thus atoms have large empty spaces in them, and the relative impenetrability of matter must have some other explanation.

It also follows from the good agreement between the measured scattering and the theoretical analysis that the force between the alpha particles and the nucleus is predominantly the electrical repulsion due to their like charges. Actually, this is true for the conditions under which Rutherford's experiments were done, in which the alpha particles did not pass through the nucleus. We shall see later that when alpha particles or other nuclei pass through one another there are other forces that act between them.

The experiments on alpha particle scattering give little informa-

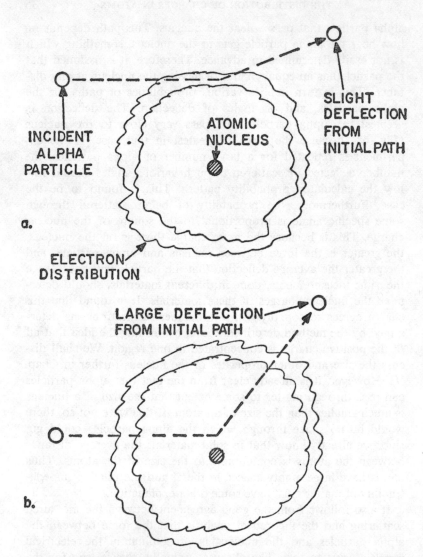

INCIDENT ALPHA PARTICLE

ATOMIC NUCLEUS

SLIGHT DEFLECTION FROM INITIAL PATH

a.

ELECTRON DISTRIBUTION

LARGE DEFLECTION FROM INITIAL PATH

b.

FIGURE 6. Deflection of alpha particles by an atom. An alpha particle that passes far from the nucleus of an atom as in a. of the figure will be only slightly deflected from its path by the light electrons. An alpha particle that passes near the nucleus, as in b., will be widely deflected, because of the greater electrical forces exerted by the nucleus, and because the large nuclear mass allows more transfer of the available momentum to the nucleus.

tion about the distribution of the electrons in atoms because the electrons, being so light, are essentially brushed aside by the heavy alpha particles as they go by. Information about where the electrons are located in atoms is given by measurements of the interaction of low-energy electron beams, or light beams, with atoms, and will be discussed later, after we consider what to expect about the electron distribution in atoms on theoretical grounds. High-energy electrons, such as in beta rays, also are scattered much more by the atomic nucleus than by the electrons, because the nucleus is so massive and because its charge is greater than that of each individual atomic electron. Measurements of scattering similar to these we have described for alpha particles have been carried out for beta-ray electrons, and those measurements also agree with the theory of scattering due to electrical forces, indicating again that, at the distances involved in these experiments, which can be as small as 10^{-12} cm, the forces between electrons and nuclei are mainly or entirely electrical. These results are important, as they indicate that similar electrical forces must act between the nucleus and the electrons within the atom, which on the average are at least that distance from the nucleus.

III

How Atoms and Their Components Behave

As we have learned, atoms are complex objects, composed of electrically charged components. To understand why and how they stay together under ordinary conditions, we must examine the electrical forces that act between the nucleus and the electrons of an individual atom. From the experiments of Rutherford described in Chap. II, it is known that this force is the electrical attraction acting between opposite charges, which acts along the line joining the charges, and decreases inversely as the square of the distance between them. This law, originally discovered by Charles Coulomb in the eighteenth century to express the force between charged pith balls, describes accurately the force between two nuclei, or between nuclei and electrons. In a many-electron atom, there is also a force between the electrons which repels the electrons from each other. But because each electron has a charge that is only a fraction of the charge of the nucleus, this force between electrons is usually less important than the force exerted by the nucleus.

Let us consider first the simplest atom, hydrogen, which has a nucleus, and one electron. A stable hydrogen atom can only exist if there is some internal motion, since if the electron and nucleus were at rest, the electric force would cause them to move toward one another, and collide in a matter of 10^{-16} sec. It would be possible to arrange the electrons and nucleus in a complex atom so that there is no internal motion, and all the forces balance out. But this equilibrium is unstable—somewhat like a long pointer balanced on end—and any outside influence such as another nearby atom, will destroy the equilibrium; the atom will either fall apart or the electrons and nucleus will come together. The need for internal motion is inescapable in all atoms.

An obvious source of this motion might be a closed orbit of the electron around the nucleus, analogous to the planetary motion in the solar system. We know that a planet moving around the Sun is a stable system that can continue indefinitely, and we might expect that the electrons revolving around a nucleus would behave similarly. Rutherford's scattering experiments led him to propose just such a planetary model of atoms. In Rutherford's model, which is in many ways the present one, the positive charge is concentrated in the center of the atom, in a small region, which must be a very small fraction of the volume of the atom. The negative electrons are away from the nucleus, revolving about it in various orbits whose average distance is the approximate size of the atom, or for some electrons, somewhat less.

However, this analogy between atoms and planetary systems when carried further, leads to unacceptable consequences, which could only be avoided by a thorough revision of the laws describing how atoms and their components move and behave. These new laws were worked out in a series of steps during the period from 1913 to 1927, and were labelled as quantum mechanics, to distinguish them from the laws of Newtonian mechanics, which govern the motion of ordinary objects. The planetary model is consistent with the properties that we know atoms have only because electrons follow quantum mechanics, rather than Newtonian mechanics.

The main discrepancy between a planetary model based on Newtonian mechanics and the known properties of atoms is that according to the model an atom of any type could occur in an infinite number of different forms, corresponding to electron orbits that are at any arbitrary distance from the nucleus, and that are arbitrarily eccentric, or deviant from a circle. In the actual solar system, the planets occupy only nine of these infinite number of possible orbits, but the comets and asteroids occupy others, and man-made satellites many more. There is little reason to doubt that we could put a satellite into any orbit around the Earth or Sun that we choose, so that all of the infinitely many possible orbits in the solar system seem really accessible.

On the other hand, this cannot be the case for the electrons in atoms. For example, if the electron in a hydrogen atom could be found in many different orbits, hydrogen atoms in different times and places would show quite varied properties. For example, hydrogen atoms with the electron in different orbits would undergo varied chemical reactions, because the forces between atoms depend on the electron orbits, and these forces influence how chemical reactions occur. Furthermore, the electron orbits that are nearer to the nucleus would on the average have lower energy than those far away, just as the orbit of a satellite comes closer to the Earth as it loses energy through air resistance. As a result, there would be a tendency for electrons in the more distant orbits to fall into the closer orbits, emitting extra energy in the process. This could occur through the effect of other electrons in the atom, or through collisions with nearby atoms. It could also occur through the emission of light by the electron. (It is known that electrically charged objects undergoing an accelerated motion, as would electrons in orbits, radiate light or other radiation. Indeed, this is the way such things as radio waves and X rays are usually produced.) Any of these mechanisms would result in the same situation that would happen if there were no internal motion at all, that is, the electrons would eventually go into an orbit that would make them collide with the nucleus, ending the existence of the atom. The time for this to happen would be something like 10^{-9} sec, much longer than in the absence of internal motion, but still much too short compared to the billions of years atoms are known to exist.

It should be remarked that their are several reasons why a similar thing, i.e., the planets dropping into smaller and smaller orbits and eventually hitting the Sun, does not happen in the solar system. One is that the solar system is far away from any other stars, relatively much more so than the atoms in a very rarefied gas are from each other. A second reason is that the force between the planets is a very small fraction of the force the Sun exerts on each planet, and the effects of the planets on each other's motion is very small except over immense time periods. Finally, the planets have small or zero electric charges, and the rate of energy loss due to radiation is very low, or zero. As a result, the orbits of the planets in the solar system are stable for billions of years or longer. On the other hand, satellites in orbit near Earth lose so much energy through collisions with air molecules that in a matter of months they fall into orbits colliding with the Earth.

Both of the problems raised about the planetary model of atoms could be avoided if, instead of an infinite number of possible orbits with lower and lower energies, there existed for each electron an orbit of lowest energy at a minimum distance from the nucleus. There could be other possible orbits with higher energy as well, but we can imagine that, normally, the electrons in the atom are found in their minimum-energy orbits. This is called the ground state. Since there are no orbits of lower energy available, the electrons cannot radiate away any of the energy of their minimum orbits. Neither can a collision with another atom decrease the orbit energy since no lower energies exist. Such collisions can, however, raise one or more of the electrons to higher-energy orbits, in which case the atom is said to be in an excited state. After this happens, the electrons will usually radiate away the extra energy they have, and drop back or decay, to the ground-state orbits. This will happen in a short time, so that the atoms will spend most of their time in the ground state. Only at high temperatures of thousands of degrees, or in other abnormal or un-earthlike conditions in which collisions take place very often, do atoms spend much time in excited states. The fact the atoms are usually in their ground states and that the ground states of all atoms of one type are similar, explains why all atoms of one type have the same properties.

Experiments done by James Franck and Gustav Hertz around 1914 showed that not only is there a lowest-energy state of atoms, but also that the excited states exist in only a small number of the infinite set of orbits allowed by Newton's laws. In these experiments, a beam of electrons (all with about the same energy, which can be varied) is directed at a gas containing atoms of some element which are presumably in the ground state. Collisions of electrons in the beam with one in the atom will cause the beam electrons to lose energy and that in the atom to gain energy, raising the latter to an excited orbit. If there were excited orbits with energy very close to the energy of the minimum orbit, this could happen no matter what the energy of the electrons in the beam. Instead, it is found that if the beam electrons have an energy below some minimum value, none of them lose any of their energy, indicating that no atomic electrons are being excited. This is recognized by the fact that the electrons that emerge after passing through the gas have the same energy as those entering the gas rather than a lower energy. When the energy of the beam electrons is increased to a certain value—which depends on the type of atom being bombarded by them—there is a sudden change, and many of the beam electrons are found to lose all their energy by exciting atomic electrons. If the beam electrons are given slightly higher energy, they lose only part of their energy while passing through the gas. However, at some specific, still higher energy, some beam electrons again lose all of their energy to atomic electrons. As the energy of the beam electrons is increased, this process repeats itself over and again until a certain energy, depending on the atom, is reached. At this energy, or any higher energy, beam electrons can lose all of their energy to atomic electrons.

Since the energy lost by an electron in the beam must be equal to the energy gained by an atomic electron, these results show that the atomic electrons in excited orbits differ in energy from those in the ground state by discrete amounts of energy. In other words the allowed orbits for electrons in atoms do not have all possible energies, but only a restricted set of energies. This phenomenon is called the quantization of energy levels in an atom. Quantization of energy levels occurs not only for electrons in atoms, but also for the particles that make up a nucleus and in many other situations. It is therefore a very important aspect of atomic and subatomic physics, and one that requires physical laws other than those of Newton's mechanics for its explanation, since those laws allow orbits with any energy to exist. These new laws are part of the set of principles known as quantum mechanics which form the major intellectual achievement of twentieth century physics.

It should be remarked again that there is a certain value of the energy, higher than any of the discrete values corresponding to the excitation of electrons to higher orbits, above which the beam electrons can always lose energy to the atomic electrons. This is called the ionization energy of the atom, and corresponds to the amount of energy necessary to enable an atomic electron to escape from the atom entirely. Once the electron is out of the atom, its energy levels are no longer quantized and the electron can have any value of energy in a continuous range, which allows the beam electron to lose any amount of energy above the ionization energy. The fact that quantization of energy levels does not occur everywhere, must be explained by the laws of quantum mechanics.

Quantum Mechanics

In order to understand how the quantization of energy levels can occur, we must consider the two contributions to the energy of an electron in orbit. One part is the potential energy, which comes from the attractive forces between the electron and nucleus, and the repulsive forces between the various electrons. The total potential energy depends on the distances between the electrons and the nucleus and the distances between all the electrons. Because the force between the electrons and the nucleus is the most important one in an atom, this corresponding potential energy is also the most important.

The other contribution is the kinetic energy of motion, which depends only on the speed of the electron. The kinetic energy itself consists of two parts: one coming from motion toward or away from the center of the orbit, and the other from motion in a circle around the center. Of course, for certain orbits, the motion and the energy could be completely of one type or the other, but usually it is a mixture of the two. The total energy is the sum of the kinetic and potential energies. Unless the orbit is a perfect circle at a constant speed, the different contributions to the energy will vary from one point of the orbit to another (Fig. 7). But the total energy will stay constant, with the change in potential energy compensating for the change in kinetic energy. This is one example of the law of conservation of energy discussed in Chap. I.

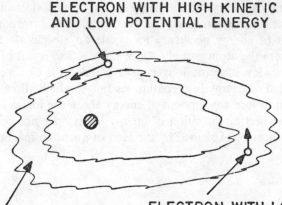

ELECTRON WITH HIGH KINETIC
AND LOW POTENTIAL ENERGY

APPROXIMATE REGION
OF MOTION OF
ELECTRON

ELECTRON WITH LOW
KINETIC AND HIGH
POTENTIAL ENERGY

FIGURE 7. Kinetic and potential energy in bound motion of an electron. An electron in an atom usually moves so that it is sometimes near the nucleus and sometimes far. When it is near the nucleus, its potential energy is very negative, and its kinetic energy is very positive, so that its speed is large. When it is far from the nucleus, its potential energy is near zero, and, by the conservation of energy, its kinetic energy is also, so that its speed is small.

The kinetic energy of an object is never negative, and is zero only when the object is at rest. The potential energy can be either positive or negative. The part of the potential energy of an electron in an atom which comes from the attraction between the nucleus and an electron is negative, and becomes more negative as the electron approaches the nucleus; it approaches zero as the electron gets very far from the nucleus. The total energy of an electron in a closed orbit is always negative, and can be proven so directly from Newton's laws in the following way. During its motion an electron with negative total energy cannot get very far from the nucleus, because if it did so its potential energy would become almost zero, while its kinetic energy would remain positive; its total energy would then be positive, which contradicts the assumption that the total energy is negative. Therefore, an electron with negative total energy must remain in an orbit near the nucleus (or else hit the nucleus). On the other hand, an electron with positive energy will be able to move very far from the nucleus, and eventually escape from the atom. Such an electron is similar to a rocket that is given enough energy by its fuel to escape from the Earth, while the bound electron is similar to a rocket that goes into a satellite orbit. The amount by which the total energy of an electron in an orbit is less than zero is called the binding energy of the electron. The energy needed to remove all of the electrons in sequence from the atom is called the total binding energy. In order to remove an electron from the atom, an amount of energy equal to or greater than its binding energy must be transferred to it. Since the different electrons in the atom will have different orbits and different binding energies, it will take different amounts of energy to remove them. The smallest such amount, corresponding to the smallest binding energy, is the ionization energy of the atom, which was measured by Franck and Hertz.

According to Newton's laws, the binding energy will depend on the average distance of the electron from the nucleus in such a way that the binding energy will increase inversely as the average distance decreases, and the average kinetic energy will adjust itself to a value equal to the binding energy. From this it follows that the observations of Franck and Hertz show that the average distance of the electron from the nucleus, and the average kinetic energy of the electron in its orbit, also are able to assume only certain values, and are quantized just as is the total energy. These conclusions are also at variance with the expectation in Newtonian mechanics, according to which the electron might have any kinetic energy, or any average distance from the nucleus.

Heisenberg's Uncertainty Relation

The idea which furnishes the key to understanding why the energy levels of bound orbits are quantized is a relationship which is not recognized in Newtonian physics involving the position and the momentum of any object. By adding this new relation to New-

a.

RANGE OF
UNCERTAINTY
OF POSITION

RANGE OF
UNCERTAINTY
OF MOMENTUM

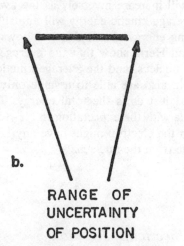

b.

RANGE OF
UNCERTAINTY
OF POSITION

RANGE OF
UNCERTAINTY
OF MOMENTUM

ton's laws, an accurate description of atomic and many subatomic phenomena can be obtained. This relation was discovered by Werner Heisenberg in the late 1920s, after a number of other partial insights had been found by Neils Bohr, Arnold Sommerfeld, and Erwin Schrödinger.

The essence of the quantum relationship between position and momentum is that the momentum of an object becomes less well determined as its position becomes better determined, and the position becomes less well determined as the momentum becomes better determined (Fig. 8). Here "determined" has two distinct meanings, one of which is the accuracy with which some quality can be measured and known, and the other the extent to which the quantity varies in its actual value. The new relation then restricts the accuracy with which position and momentum can jointly be measured, and also restricts the range of variation they may have for an electron in the atom. These two restrictions are related, and we shall see the logic of the relationship later.

More precisely, if Δx is the accuracy with which the position of an object along any line is determined, and Δp is the accuracy with which the momentum of the object along the same line is determined, the relation states that the product of Δx and Δp cannot be decreased below a certain value. Since momentum is a product of mass and velocity, at least for objects that are moving slowly compared to the speed of light, the same relationship says that the product of Δx and Δv cannot be reduced below a minimum value. Although the two forms of the relationship are equivalent, physicists generally use the one involving position and momentum, because for these quantities the minimum value for the product of Δx and Δp is the same for all objects, whatever their mass, while for position and velocity the minimum value depends on the mass of

FIGURE 8. Heisenberg's relation. The figure shows two illustrations of Heisenberg's relation. In a., the position of a particle is well determined, represented by the small blurred region. The momentum is therefore poorly determined, represented by the large number of different arrows that might represent the momentum. In b., the momentum is well determined, as indicated by the small array of arrows, and thus the position is not so well determined, as indicated by the large blur.

the object. Note that Heisenberg's relation does not restrict either Δx or Δp individually, but only their product. Either position or momentum can be determined as precisely as desired, but not both together. The minimum value that the product of Δx and Δp can have is related to a number called Planck's constant, and is equal to approximately 5×10^{-28} g cm^2/sec. Its dimensions are those of momentum multiplied by distance. This product is sometimes called action for obscure historical reasons. Planck's constant, which will be discussed later in connection with the quantum theory of light, is greater than the minimum $\Delta x \, \Delta p$ by a factor of 4π, and is equal to 6.6×10^{-27} g cm^2/sec. Physicists often use the symbol $\hbar$ for Planck's constant divided by 2π, and we shall follow that practice here.

Heisenberg's uncertainty relation, as it is called, was discovered in 1927. It immediately helped to clarify the meaning of several things that had been found earlier, but not quite understood in physical, as opposed to mathematical, terms. The discovery led Wolfgang Pauli, a leading physicist of the time, to remark that the Sun had finally risen on the quantum theory. To appreciate why Pauli felt this way, it must be realized that in Newtonian physics there is no relationship at all between the accuracy with which different quantities are determined. The momentum of an object may be known more or less well, and, independently, its position may be known more or less well, but there is no reason why both cannot be determined as accurately as desired, so that the product of Δx and Δp can become arbitrarily small. For the relatively large objects to which Newtonian physics was applied, the position and momentum were never determined extremely accurately on the scale of atomic dimensions, so a limitation of the type described by Heisenberg's relation was unnoticed. For instance, suppose a ball weighs 100 g, and we know its position with an accuracy of 10^{-3} cm, and its velocity with an accuracy of 10^{-3} cm/sec, both of which are quite difficult to achieve. The product of Δx and Δp would then be 10^{-4} g cm^2/sec. This is some 10^{23} times greater than the minimum value allowed by Heisenberg's principle. In other words, the usual limitations of accurate measurement for ordinary objects are far more serious than the restriction of Heisenberg's relation, which explains why that relation was not discovered until physicists began to study atoms and their constituents.

On the other hand, for an electron in an atom, the product of Δx and Δp may be close to the minimum allowed value ($\frac{1}{2}$ $\hbar$). The value of Δx cannot be much larger than the size of the atom, because we know that the electron is in the atom. Similarly, the value of Δp cannot be much larger than the average momentum of an electron in an orbit, because we know that the electron has about the right kinetic energy, and hence velocity and momentum, to keep it in an orbit. These values, for a typical electron, correspond to a Δx of about 10^{-8} cm, and Δv of about 10^8 cm/sec. This makes the product Δx and Δp equal to about 10^{-27} g cm/ sec, which is close to the minimum value. This suggests that Heisenberg's relation may be of great importance for the internal motion of atoms. Actually, Heisenberg's relation approximately determines the minimum size of atoms. If atoms were much smaller than they are, the position of the electrons would be more precisely known and therefore the velocities would become very imprecise. If the velocity is very imprecise, the average kinetic energy must be large, and the total energy would be positive rather than negative as it should be for a bound electron. In other words, atoms are as small as they can be without making the kinetic energy due to the uncertainty in velocity larger than the potential energy. This implies a minimum size for atoms, which can be expressed in terms of Planck's constant and of the mass and the electric charge of electrons. The value turns out to be 10^{-8} cm. So quantum theory is able to explain the numerical value of the size of atoms which previously was a purely empirical quantity.

Heisenberg's relation also implies that most of the orbits allowed by Newtonian mechanics actually do not exist in atoms. One example of this is an electron orbit that would be a perfect circle in Newtonian physics. Such a motion would correspond to $\Delta p_r = 0$, where p_r is the radial momentum, toward or away from the center. But according to Heisenberg's principle, $\Delta p_r = 0$ would require Δr to be infinite, where r is the distance from the center. But in a circular orbit Δr is actually zero, so that motion at a fixed distance from the nucleus is not a possible orbit for the electron. This statement is also true, strictly speaking, for the motion of a planet around the Sun, but in that case, the deviation from circularity required by Heisenberg's principle is so small that it is lost in other effects that give a deviation from circularity.

Another example is an elliptical orbit in which the electron passes near the nucleus, and then returns out to some maximum distance d. In Newtonian physics, the distance d can be arbitrary, and in particular d could be very small. However, such an orbit would have a Δr of approximately $\frac{1}{2}$ d. In order to satisfy Heisenberg's relation in this orbit, Δp_r would have to become very large and so the kinetic energy, which is proportional to $(\Delta p_r)^2$, would become too great for the electron to remain bound. This implies that orbits at a small distance from the nucleus, and hence with large binding energy, are also forbidden to electrons. On the other hand, electrons with positive energy are usually very far from the nucleus and so their position need not be well defined. Therefore, the Heisenberg relation does not restrict Δp and there is no restriction on the possible energies of such electrons.

If the coordinate is confined to a limited range, as, for example, for an electron in a box, the Heisenberg relation allows the momentum to be quantized, i.e., to assume only a limited set of values rather than the arbitrary values allowed in Newtonian mechanics. One important application of this relation is to the quantity known as *angular momentum,* which is related to angular velocity, or motion about the center, in the same way as radial momentum is to radial velocity, or motion toward the center; that is, through multiplication by the mass of the moving object (Fig. 9). The angular coordinate is simply the angle between a fixed direction and the line from the center of the orbit to the object. This coordinate, by its definition, is always restricted in values from $0°$ to $360°$. The Heisenberg relation then allows the angular momentum to assume only a limited number of values, which turn out to be integer multiples of $\hbar$, that is $0\,\hbar$, $1\,\hbar$, $2\,\hbar$, etc. This quantization of angular momentum was first discovered by Bohr in 1913, long before Heisenberg's principle gave a reason for it, and was used by Bohr to explain why in certain cases the energy levels of hydrogen atoms were quantized. Bohr actually obtained fairly accurate values for the quantized energy values in hydrogen, indicating that the quantization of angular momentum is probably correct. The allowed orbits must have the quantized values of angular momentum.

A more direct proof of this was first found by Otto Stern and Walter Gerlach. They passed a beam of atoms through a magnetic

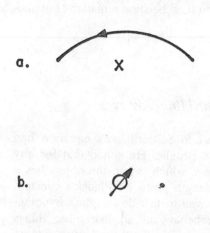

FIGURE 9. Two kinds of angular momentum. An object can have either or both of two kinds of angular momentum. If it is moving in some curved path, or orbit, it has an orbital angular momentum relative to any point such as X, that is off the orbit. The orbital angular momentum has a direction perpendicular to the plane of motion. This situation is shown in a. If the object is rotating on an axis passing through it, it has a spin angular momentum. The direction of the spin is that of the axis of rotation, as in b.

force that varied from point to point in space. According to Newtonian physics, the atoms in the beam should be deflected by various amounts, proportional to the angular momentum of the electrons in their orbits. If the angular momentum was not quantized, the amount of deflection could have any value. Instead, it was found that the atoms are only deflected by a few distinct amounts corresponding to the quantized values of the angular momentum. So here, too, a correct description was found by supplementing Newtonian mechanics by the requirements of Heisenberg's principle.

These results begin to indicate how the Heisenberg relation acts to supplement the equations of Newtonian physics by allowing the electrons to have only certain energies of the infinite number that Newtonian physics allows. The precise values of the allowed energies can be calculated for simple atoms such as hydrogen using a more detailed mathematical treatment of the equations of Newtonian physics combined with a mathematically precise statement of Heisenberg's relation. For more complex atoms, the calculation of the allowed energies is more difficult, but is still possible through the use of an equation discovered by Erwin Schrödinger in 1926, about the time the Heisenberg relation was discovered. While *Schrödinger's equation* seemed to have a different basis than Heisenberg's relation, it was soon recognized by Schrödinger,

Dirac, and others, that these were alternative ways of expressing the same facts, and the one to use then became a matter of mathematical convenience.

De Broglie–Schrödinger Waves

The basic assumption that led to Schrödinger's equation had been made in 1923 by Louis de Broglie. He stated that for any object there is an associated wave, which, when the object has a definite momentum, has a wavelength given by Planck's constant divided by the momentum. One way to test de Broglie's hypothesis is to see whether electrons behave at all like other things known to have wave properties, such as light or X rays. For an electron that is not part of an atom and is moving freely through space, the kinetic energy and the amount of momentum can both be expressed in terms of the velocity, and therefore the amount of momentum can be determined if the energy is known. Since the amount of momentum determines the wavelength of the wave associated with an electron, this means that if we know the energy of an electron we also know the associated wavelength. It turns out that for an electron whose energy is 200 eV, the wavelength is 10^{-8} cm, or about the same as the wavelength of a typical X ray. It is known that when X rays pass through certain crystals, they undergo *diffraction,* that is, they are deviated by the crystal from a straight-line motion. Another example of diffraction is that light going through a small hole will not give a simple image of the hole, but instead gives a series of light and dark rings concentric with the central image (see insert, Plate 1). It is expected that if electrons have wave properties, they will be diffracted by crystals in a way similar to X rays of the same wavelength even though the mode of interaction between electron and the individual atoms in the crystal is not similar to that between the X ray and the atoms. This is indeed what happens, and this was first observed by Clinton Davisson and Lester Germer, and concurrently by George Thomson in 1927 (Plate 2). It is harder to observe this type of diffraction of other particles, because the wavelengths are usually so short that even the close spacing of atoms in a crystal does not give much diffraction for them. However, the kind of particle called neutrons can be made with very low momentum, and hence large wavelengths, and for these slow neutrons diffraction by crystals is also observable. For other objects, the wave properties must be observed by more indirect means, but physicists are now convinced that all subatomic particles have waves associated with them, and that the wavelengths satisfy the relations assumed by de Broglie. The de Broglie waves also display another phenomenon, already well known for light waves, called interference. Interference occurs when a light wave is split into two parts and then recombined at various points; the intensity or amount of light reaching different points is not just the sum of the intensity contained in one part and the intensity contained in the other. Instead, the intensity in the combined beams can vary between zero and twice the sum of the separate intensities, depending on the conditions of separation and recombination.

Thus far we have discussed the de Broglie waves for a particle moving freely through space, with no forces acting on it. In that case, the momentum remains constant, and so does the wavelength. For a particle with forces acting on it, such as an electron in an atom, the momentum varies from place to place as the force varies, and hence the wavelength will also vary from place to place. Situations like this also occur for light waves moving through a material whose density varies from point to point. In such a case, it had been known for a long time that the motion of a light wave is determined by an equation which describes how the wavelength of the light varies from point to point. Schrödinger used an analogy with the behavior of light to write an equation which combines the relation between momentum and wavelength with the law of conservation of total energy. As we have seen, this law relates the velocity or momentum contained in the kinetic energy and the position contained in the potential energy. The resulting equation explains how de Broglie's wave varies from point to point in space in an atom, or wherever else forces act on particles and also how the wave varies in time. When this equation is solved for an electron in an atom, it is found that physically sensible solutions exist only for certain values of the total energy. It is these energies that are the ones allowed for the electrons, and which correspond to the values measured in the Franck–Hertz type of experiment.

The Schrödinger waves describing an electron in an atom can be thought of as containing a range of different wavelengths, rather than the single wavelength of a simple wave form. They are similar in this respect to the sound waves produced by a complex musical instrument. The Schrödinger wave equation can also be interpreted as describing precisely the combination of wavelengths which occur for an electron in an atom. A wave with such a combination of wavelengths may exhibit the property of being small except in a certain region of space, unlike a simple wave, which extends throughout space. The localized type of wave is often called a wave packet. We shall use the word "wave" to mean any combination of wavelengths, rather than a simple wave form. Because of the de Broglie relation between wavelength and momentum, the wave of an electron in an atom will contain many different momenta. This range of momenta is another aspect of Heisenberg's relation. Since the electron has an approximately well defined position—just from the fact that it is in the atom and not far from the nucleus—according to Heisenberg's relation it cannot have a definite momentum. The range of momenta Δp contained in the atomic wave packet obtained from the Schrödinger equation, and the approximate accuracy with which the position is known, which may be taken as the size of the atom, have a product approximately equal to $\hbar$. This result again shows that the electron's motion in the atom adjusts itself to the requirement of Heisenberg's relation, and that this relation determines the approximate size of the atom.

An obvious question about the de Broglie–Schrödinger waves is what they are made of, or put differently, what substance the wave is describing the motion of? Such a question was also posed in the nineteenth century, when light waves and other electromagnetic waves were discovered. In that case, physicists initially considered the waves to be vibrations of a mysterious substance called ether, and made various mechanical models of its motion. But eventually such ideas proved untenable, and the waves were regarded as independent things, as real as the atoms with which they interact. In the case of electron waves, Schrödinger at first suggested that the waves actually represented the motion of the electron's charge, so that in an atom, the charge of the electrons was physically distributed over the region of space in which the wave packet is different from zero. This interpretation did not last very long, as it was shown by others that in some circumstances an electron wave packet that begins as a distribution localized in a small region of space soon changes into a different wave packet which is spread over a much wider region in space. Under Schrödinger's interpretation, this meant that the electron's charge was being dissipated and eventually there would be no chance of finding much of the electron anywhere. However, this behavior is never observed. Instead, whenever the electron is found, its total charge is localized in a small region of space. Hence a direct physical interpretation of the changing wave packet is unreasonable.

The correct interpretation of the de Broglie–Schrödinger waves is much more subtle, and more surprising in terms of Newtonian mechanics. These waves describe the probability that measurements of the position, momentum, or other qualities of the electron, will obtain certain values. For example, the *wave function,* or magnitude of the wave, describing an electron in an atom has the property that it is relatively large near the nucleus, and small far from the nucleus. According to this probability interpretation, this means that if a very accurate measurement of the position of an electron is taken, there is a high probability of finding it near the nucleus, and a small probability of finding it far away. But in either case the measurement will detect a whole electron, not a partial one, as Schrödinger's original interpretation would imply.

The probability distribution of electron positions in an atom does not vary with time, provided that the atom has a definite energy. The distribution is different for the different energy levels of the atom. However, since the atom is usually in its lowest energy level, it is the distribution corresponding to that level which is usually relevant. One might think of this distribution as a kind of time average of the positions that the electron could occupy in a classical orbit. Since the electron's angular velocity is very large according to Newtonian physics, any measurement carried out over an extended period of time would give such an average over the orbit anyway. But this image gives a somewhat distorted picture for several reasons. There is no way of telling whether electrons follow orbits in the atom, and no reason to believe they do. In order to follow a classical orbit, the electron would, at any time, have to have a definite value for its exact velocity and its exact position. However, according to Heisenberg's relation, there is no way of determining the exact position and the exact velocity at the same time. Any measurements taken in order to follow the orbit of the electron would disturb its motion so greatly that we could not determine what orbit it was following. Therefore, it makes no sense to ascribe definite orbits to the electrons. This conclusion is based on the fact that for electrons in an atom the size of the atom, and hence the properties of what would be the orbit, are determined by the Heisenberg relation. The same is true for the particles in a nucleus, where again the size of the nucleus is largely determined by Heisenberg's relation. For other phenomena, such as the motion of the Earth and Moon, the size of the system is much larger than the minimum size required by Heisenberg's relation and the concept of orbits can be applied to the motion with a high degree of accuracy.

Quantum States

The concept in quantum mechanics that replaces that of a definite orbit in Newtonian mechanics is that of a *state*. A physical system is said to be in a definite state when it is known to have values for a set of quantities which can all be determined together without contradicting Heisenberg's relation. For example, a particle with values for each component of its momentum is in a definite state. For any physical system whose constituents we know, and in which we know the forces that act among the constituents, we can in principle calculate all the possible states of the system. Also, we can calculate, from the Schrödinger wave equation, how the state of the system changes in time. This equation therefore plays the role in quantum mechanics that Newton's law does in Newtonian mechanics. When referring to an electron in an atom, we should describe it as being in one or another state, rather than in some orbit, as the latter is not really meaningful. However, there is in many cases a rough correlation between states and orbits that is loosely defined, in the sense that the electron has some of the properties such as energy and angular momentum, that it would have in an orbit. Physicists therefore sometimes use the Newtonian language to describe such situations.

It is possible to measure the probability distribution of positions for an electron in some state of the atom without trying to follow the electron in an orbit. This can be done by employing the method used by Rutherford to discover the nucleus. A beam of electrons is directed at the atom, and the angular deflection of the electrons in the beam is measured. Each beam electron will be deflected by a different amount. The individual deflection is unpredictable, as it depends on the exact electron trajectory and the exact electron orbits in the atom, which are indeterminate. However, the result of a large number of individual deflections follows a simple probability law, and depends on the distribution of positions of the electrons in the atom, which scatter the beam electrons. Hence by measuring the pattern of deflections of the beam, an average distribution of all the electrons in the atom, each of which helps to scatter the beam electrons, can be inferred. Obtaining information about the individual electrons in a multiple electron atom raises other questions which will be discussed later.

There are also other measurable properties of the atom that depend on various aspects of the distribution of its electrons. The forces that atoms exert upon one another, which govern the chemical properties of the substance made up of the atoms, depend on the electron probability distributions of each atom. So does the way in which the atom responds when radiation hits it. Indeed, physicists have learned that all of the things that we can measure or observe about the electrons in atoms are determined, at least in the sense of probabilities, by the Schrödinger waves associated with the atom. If the relative amounts of various wavelengths that are present in this wave are known, then all the properties of the atom can be predicted. For simple atoms, the Schrödinger wave can be calculated directly from the Schrödinger equation. For hydrogen and helium atoms, this has been done with very great accuracy, and the properties of these atoms are well understood. For atoms with many electrons, it is harder to solve the equations because of the forces exerted by the electrons on one another. In these cases, physicists have used various approximations for solving the equation, and have obtained substantial information about the properties of the atoms in that way.

Quantum Numbers and the Exclusion Principle

In order to understand all the properties that atoms are observed to have, it is necessary to ascribe certain characteristics to electrons beyond those of Newtonian mechanics, and in addition to the Heisenberg relation. We have seen that Heisenberg's relation accounts for the fact that each atom has a minimum-energy state that an electron can occupy. However, not all of the electrons in a particular atom are found in this state. For example, in the lithium atom, which has three electrons, two are ordinarily found in the minimum-energy state, but the third occupies a higher-energy state. No matter how long one waits, the third electron never drops to the lowest state that the other two are in. It might be thought that the electric repulsion between the electrons has something to do with this, but that is unlikely for two reasons: first, two electrons do have the same energy, and the repulsion acts between them, and furthermore, the same behavior is exhibited in heavy atoms with many protons in the nucleus, and many electrons. If the repulsion of the electrons were what kept most of the electrons out of the lowest-energy orbit, we would expect that in heavy atoms more electrons would drop to the lowest orbit, since the attraction of the nucleus is relatively more important than the repulsion of two electrons. Instead, even in a uranium atom, which has ninety-two electrons, only two are ever found in the lowest-energy state. This can be proven by studying the absorption of X rays by uranium atoms. This absorption is different for electrons in different states, and the amount of absorption depends on the number of electrons in the state. By careful measurements of the amount of absorption of X rays of various energies, the number of electrons with each energy can be determined. These numbers are characteristic of each type of atom, and are an important fact to be accounted for by atomic theory.

Another relevant observation is that if one of the electrons is removed from a low-energy orbit, say by absorbing X rays with enough energy to remove the electron from the atom, one of the electrons in a high-energy state will rapidly drop into the place of the electron that has been removed, but no other electrons will follow this one. It is as if only the presence of two electrons in the orbit were keeping the others out; if one electron is removed, one, and only one, electron can replace it, effectively barring any further intrusion. Strictly speaking, this is also true in reverse, although not as easy to demonstrate. If an atom already has the maximum number of electrons allowable in a particular high-energy orbit, say two, then it is impossible to bring an electron from a lower-energy state into this higher-energy one, say by having the lower-energy electron collide with an electron in an external beam, as in the Franck–Hertz experiment. No matter what the process involved, electrons behave as if their presence or absence in a given orbit influences the availability of that orbit to other electrons.

It is important to recognize that electrons influence only each other this way, and have no such effect on other particles. For example, there is another negatively charged particle, called the *muon,* which we will discuss in more detail later. Muons can also go into orbits together with electrons around nuclei. However, a muon can enter into a given orbit regardless of whether that orbit already contains electrons. It is generally believed that muons would affect other muons in orbit in the same way that electrons affect each other, but atoms containing several muons have never been studied, because muons are difficult to produce and disappear quickly.

All of these facts can be summarized in a single principle, called the *exclusion principle,* or sometimes the Pauli principle, after the physicist who discovered it in 1925. In order to state the exclusion principle, we must specify which attributes describe an electron as completely as Heisenberg's relation allows. First consider an electron in space, far away from other bodies. In Newtonian mechanics, such an object could be completely described by the three numbers that measure its distance along three reference lines from some fixed point and three more numbers measuring its momentum along the same three lines. These numbers are the components of position and components of momentum. Heisenberg's relation tells us that the positions and momenta of an electron cannot be determined simultaneously. Instead, we can determine all three positions, or all three momenta, or other combinations of three numbers. With a certain addition we shall mention below, these three numbers give as complete a description as possible of a single electron in free space. Usually, the three numbers giving the momenta are used for a free electron, because these remain constant with time, while the positions change. For a single electron in an atom, we could use the same three numbers, but it is more convenient to use numbers that would be constant in time for that case. One of these numbers could be the energy of the electron. The energy is usually insufficient by itself to specify the state, for reasons to be discussed below. Two more numbers that are constant in time, and that can be specified together with the energy are the magnitude of the angular momentum of the orbit, and the projection or component of this angular momentum along some line. No other quantities describing the orbit can be determined together with these three, although others could be substituted for one of the three. Any set of quantities describing an electron, or other particle, with the property that they can be determined together and no other quantity can be determined in addition is called a complete set of *quantum numbers*.

It might be thought that the projection of angular momentum along two distinct directions could be specified together, as can be done for linear momentum. However, angular motions around different directions are not quite independent, as are linear motions along different directions. For instance, starting at one point, angular rotations of 180° around each of two perpendicular directions will lead to the same point in space, while two linear motions along different directions never can do this. This interdependence of angular motions leads, together with Heisenberg's relation, to the impossibility of specifying more than the magnitude of the angular momentum, and its projection along any one axis. It is possible to prove that for a given value of the magnitude of the angular momentum, say J times $\hbar$, the projection along any axis is quantized and can take on the values, $J\hbar$, $J-1$ times $\hbar$, $J-2$ times $\hbar$, down to $-J$ times $\hbar$. For instance, if $J=1$, the angular momentum projection could be $\hbar$, 0 or $-\hbar$, along any axis. The number of distinct values of the angular momentum projection is therefore $2J+1$, or three values when $J=1$. Of course, when $J=0$, the projection must also be zero, because the value of the projection along any line cannot be larger than the total.

While no other quantities involving the electrons' motion through space can be determined together with the components of momentum—or with the set of energy, total orbital angular momentum, and projection of angular momentum—there is yet another quantity that must be added to these to make a complete set. While the electron revolves about the nucleus, it also rotates about an axis passing through itself just as the Earth rotates on its axis while it revolves about the Sun. This rotation of the electron produces an additional angular momentum, called *spin* (see Fig. 9). As with orbital angular momentum, the spin is also quantized. However, for an electron, the magnitude of the spin always has the value $\frac{1}{2} \hbar$, unlike the orbital angular momentum, which can be any integral multiple of $\hbar$, or even zero. The direction of the spin can vary, and the projection of the spin along a reference line can be $+\frac{1}{2} \hbar$ or $-\frac{1}{2} \hbar$, corresponding to its spinning parallel or antiparallel to the directed line. This projection of the electron spin along a reference line can be specified along with any of the sets of three quantities discussed above to form a complete set of quantum numbers to describe an electron. The projection of the spin along two directions cannot be specified together, for the same reason that was explained in the case of two distinct projections of orbital angular momentum. Other particles also have spins, which are an integer times $\hbar$, or a half-interger times $\hbar$, and these will be discussed later. Any particle with specific values of a complete set of quantum numbers is in a definite state. Since the complete set may involve spin as well as orbital motion, I shall use the phrase orbital state to refer just to the quantum numbers of orbital motion, i.e., the orbital angular momentum, and the energy.

The spin has effects other than to enter into the quantities that make up a complete set. A rotating charge produces a magnetic force that can act on either another rotating charge or on a charge moving through space. Such magnetic forces are also produced by orbital motions. However, the orbital magnetism vanishes if the orbit has no angular momentum, as in the ground states of atoms such as hydrogen or sodium. In this case, only the spin magnetism remains. These magnetic forces, or their corresponding potential energies, influence the properties of atoms in various ways. For example, if an atom is near a magnet, the energies of electrons with different spin projections change from what they are when the atom is isolated, and become unequal to one another. Such energy shifts result in changes of the wavelength of light emitted and absorbed by the atom. The observed wavelengths agree accurately with the predictions based on the assumption that the electron is spinning, and that the spin is influenced by the magnet.

The Exclusion Principle

The content of the exclusion principle is that no two electrons in the universe can be in the same state, that is, can have the same values for all of a complete set of quantum numbers. For example, if two electrons are in free space and have the same three components of momentum, their spin projections will have opposite values. A third electron cannot have the same momentum as the first two, since its spin projection would have to agree with one of the other two, which would contradict the exclusion principle. Similarly, in an atom containing many electrons, no two electrons can have exactly the same value of energy, magnitude of orbital angular momentum, orbital angular momentum projection, and spin projection, although they may agree in any three of these. For example, two electrons can be in the same orbital state, but they must have opposite spin projections. The validity of the exclusion principle implies that all electrons must be identical in their properties, at least insofar as these properties can influence the quantum states of an electron. If there were several types of electrons, differing slightly in some property such as charge or mass, it would not be possible for the exclusion principle to be satisfied. If two such electrons were in different states at one time, there would at any later time be some probability of finding both electrons in the same state because the difference in properties would cause the state of each electron to evolve somewhat differently in time. Only if all observable properties are identical can this problem be avoided. This conclusion has an interesting consequence for an old philosophical idea known as the identity of indiscernibles. According to this idea no two objects can be identical in all their properties other than their location in space. There is a story in the writings of Leibniz, a philosopher who believed in this idea, to the effect that the Princess Sophia once defied a nonbeliever of the principle to find two leaves in her garden that were identical, but the man was unable to do so. Leibniz then concluded that the principle rules out the existence of atoms. Leibniz' nonbeliever would have been more successful had he searched for identical electrons. We may also reverse the argument and conclude that the identity of all electrons disproves the principle of identity of indiscernibles.

The seemingly innocent requirements of the exclusion principle have tremendous implications for the properties of matter, both of atoms and for electrons not bound in atoms. These implications are relatively independent of the forces that act between electrons, and are, like Heisenberg's relation, rather specific to quantum mechanics. For example, the impenetrability of matter is more a consequence of the exclusion principle than of forces exerted between atoms. When two atoms are moved near one another, the electrons in one cannot occupy the orbits around the other nucleus that are already occupied by the electrons in the other atom. Therefore, the atoms cannot easily overlap. This resistance to the overlap of atoms is exhibited in the resistance to compression by a solid body and the resistance to penetration of one solid by another. Thus the exclusion principle acting between electrons gives atoms a semblance of solidity even though there is very little mass in most of the volume of an atom.

Another important consequence of the exclusion principle is the periodic system of chemical elements. We consider the series of chemical elements beginning with hydrogen. Each element differs from the preceding one by having one extra positive charge in the nucleus, and one extra electron. The object of our inquiry is to understand what the arrangement of the electrons will be in the lowest-energy state of each of these elements. For the one electron in hydrogen, the answer is simply that the electron will occupy the lowest orbit allowed by Heisenberg's relation. This orbit has a binding energy of about 13.6 eV. It has an orbital angular momentum of zero, and an orbital angular momentum projection of zero. Of course the electron may have either $+\frac{1}{2}\,\hbar$ or $-\frac{1}{2}\,\hbar$ for its spin projection. In ordinary hydrogen, either of these two spin projections can occur with equal probability. It is also possible to produce hydrogen ions containing two electrons and a hydrogen nucleus. These ions have both electrons in the lowest-energy state, modified slightly by the repulsion of the electrons. The spin projection of the two electrons have opposite values in this hydrogen ion. Such negative hydrogen ions play some role in the chemistry of hydrogen.

Next consider the helium atom which has two electrons. The lowest-energy level of helium corresponds to both electrons in an orbit similar to that described above for hydrogen, with the electrons having opposite spin projections. Since these two spin projections represent components of angular momentum along the same line, they can be added, as was done for linear momentum components in Fig. 3. Because the spin components have equal magnitude and opposite sign, they cancel each other, and the total spin of this state of helium is zero. This prediction is borne out by careful measurements of the magnetic properties of helium gas, which are quite different for spinning or nonspinning levels.

Unlike the case of hydrogen, it is not possible to form negative helium ions by adding a third electron. The lowest-energy orbit has zero orbital angular momentum, so that any electrons in this orbit must differ in their spin projections. Therefore, only two electrons can have this lowest energy. A third electron with this energy would necessarily agree with one of the first two in its other quantum numbers, which contradicts the exclusion principle. Therefore, the third electron must occupy a higher-energy orbit. This higher orbit is further away from the nucleus, and the extra electron behaves as if it were influenced only by the neutral helium atom. As the atom exerts only very small forces on the electron insufficient to bind it the electron easily escapes this atom, and hence the negative ion cannot exist for any length of time. This fact helps account for why helium cannot form chemical compounds with metals, which readily give electrons to other atoms. In addition helium doesn't form compounds by losing an electron to another atom, because its two electrons are very tightly bound, and another atom cannot exert enough force to pull the electron away from helium. These properties of helium are sometimes expressed by the phrase that the electrons in helium form a closed shell, meaning that some properties of all the electrons—such as total spin—tend to cancel each other, and that electrons are not easily added to the existing configuration. A closed shell occurs whenever an atom contains the maximum number of electrons of a particular energy allowed by the Pauli principle. Atoms with closed electron shells are magnetically and chemically rather inert. We shall see that electron shells also occur for atoms other than helium.

The next element is lithium, which has a nuclear charge of three and three electrons. Again, the lowest-energy orbit has zero orbital angular momentum and can only contain two electrons, so the third electron occupies a higher-energy orbit. In lithium, this electron is bound because the greater nuclear charge means that this third electron "sees" an object with a total charge of one, consisting of the nucleus and the two inner electrons. However, the electron binding energy is small, compared say to that in hydrogen, and the electron is easily detached, making lithium a typical metallic element that forms compounds by giving electrons to other atoms.

The next element beyond lithium is beryllium, with a nuclear charge of four and four electrons. The fourth electron can go into the same orbit as the third by having the opposite spin projection. We might assume that beryllium would be a closed shell. However, there is another set of orbits with about the same energy as the outer orbit in beryllium, but with angular momentum of $1 \hbar$. Electrons can occupy these next orbits with little change of energy, and therefore beryllium does not behave like a closed-shell element. Instead, the next six elements beyond beryllium can each add an extra electron in the new orbit, until the shell finally closes at the element with ten electrons—neon.

This process of filling shells is repeated again and again as electrons are added. Each time, the element with the closed shell is chemically inert, whereas the element just after the closed shell is a chemically active metal that easily loses electrons. Those elements that have several electrons more than a closed shell lose electrons less easily because the electrons are attracted more strongly by the greater nuclear charge. Indeed, when the atom is near the next closed shell, it becomes more likely for it to gain electrons than to lose them when another atom approaches, and the element has the chemical properties of a nonmetal like oxygen.

The precise number of electrons allowed in each closed shell is an important quantity in view of its direct connection with the chemical properties of various atoms. Our discussion suggests that this number is simply the number of orbits having exactly or approximately the same energy, but differing in at least one of the other three quantities, total orbital angular momentum, projection of orbital angular momentum, or projection of spin. This number of orbits is at least partly determined by another very general principle of atomic and subatomic physics called the principle of rotational invariance. This principle states that for an isolated system, such as a single atom, the energy does not change if the system is rotated through a fixed angle about any axis. One way to accomplish such a rotation without disturbing the system would be for the observer to rotate in the opposite direction through the same angle. For this rotated observer, the system appears to be rotated by the amount desired. The principle of rotational invariance can be proven directly if the energy is the sum of the kinetic energy and electric potential energies. It is also valid for all other types of energy we know, and is therefore assumed to be true for types of potential energy yet undiscovered.

An important consequence of this principle is that all orbits with the same magnitude of angular momentum, but with different angular momentum projection along some axis, have the same energy. This is because the orbits with different values of the projection of angular momentum can all be changed into one another by rotations around a line perpendicular to the direction of projection. For example, the orbit with projection $-J\hbar$ can be changed into the orbit with projection $+J\hbar$ by rotating through 180° around a perpendicular axis. Since the rotation does not affect the energy of the orbit, these orbits must all have the same energy. Two states with the same energy, but different values of some other quantum number are called *degenerate states*. This is a simple example of an application of the invariance principle to discover relations between different states. Such applications are very common in contemporary physics, and we will be using them again and again when we discuss nuclear and particle physics.

Since we know that there are $2J+1$ orbits corresponding to different angular momentum projections of the same angular momentum, $J\hbar$, there will be at least $2J+1$ orbits of the same energy. Furthermore, under many circumstances, the rotational invariance principle implies that electrons with either of the two possible spin projections and equal orbital angular momentum also have the same energy. In these cases, the number of different states having the same energy, and possessing a value $J\hbar$ for the total angular momentum is $2(2J+1)$. The reason this is not universally true is that the spin and orbital angular momentum influence one another slightly, especially when the electron is near a nucleus with a large positive charge. When this happens, the spin and orbital angular momentum cannot be assigned independently to the electron, but instead become combined into a total angular momentum $J_T\hbar$, and the number of different orbits with the same energy is $2J_T+1$. This effect, however, is not important in atoms with small nuclear charge.

The remaining factor that determines the number of electrons in a closed shell is that it is possible to have orbits with the same energy but different magnitudes of orbital angular momentum. This does not follow from rotational invariance, because orbits with different total orbital angular momentum cannot be rotated into each other. It is instead a consequence of the fact that the main force acting on the electrons is the electrical attraction of the nucleus. To the extent that this force determines the energy of the orbits, it is possible to have several orbits with different magnitudes of orbital angular momentum but the same energy. This does not happen for the lowest-energy orbit, which always has an orbital angular momentum of zero. As we have seen, there are two states corresponding to this energy, and the exclusion principle allows a maximum of two electrons in this orbit and the first closed shell occurs with helium. The next higher orbits can have an angular momentum of zero or one, with approximately the same energy. According to the discussion above, there are two and six such states, respectively, and by the exclusion principle, only one electron may occupy each state for a total of eight with the same energy. This implies that the second closed shell occurs at a total of ten electrons in the element neon, which is chemically and physically very similar to helium, because further electrons would again have to occupy states with much higher energy. Again, the next higher-energy orbits can have an angular momentum of zero or one, and so again can contain up to eight electrons, leading to another closed shell at the element with eighteen electrons which is argon, a gas composing 1% of the Earth's atmosphere, and again similar to helium. Beyond argon, it becomes harder to predict the orbital angular momenta that are associated with a particular energy because the effects of the electrons upon each other become important. In heavier atoms—those with more electrons—the determination of which elements have closed shells and of the chemical properties of various elements is a more complex matter, although some regularities have been found to exist.

The application of quantum mechanics to the Rutherford model of the atom has therefore proven capable of giving a detailed and accurate picture of the systematic chemical and physical properties of atoms. It has also been possible to understand the interactions of atoms with each other to form molecules using quantum mechanics. I cannot discuss this topic in detail here, but it is not inappropriate to quote a statement of Paul Dirac, who in 1928 described the Schrödinger equation for the many electrons in an atom as "including all of chemistry, and most of physics." The latter half of this statement has not remained true, because of new discoveries, but the former half has become truer as time passes.

The Reasons for Heisenberg's Relation

Heisenberg's relation is of great importance for understanding not only the structure of atoms, but also many other aspects of atomic and subatomic physics. In some sense, it represents the fundamental difference between Newtonian and quantum physics. The question then arises as to how there can be a restriction on the possible determination of quantities that were previously thought to be completely determinable. Of course, inaccuracies in measurement are always present, and often so great that the product of Δx and Δp is much greater than the minimum that Heisenberg's relation requires. Yet it seemed plausible that these inaccuracies could be reduced as much as desired, so that their product could be made as small as desired. But Heisenberg's relation suggests that while either position or momentum can be separately determined as accurately as desired, the process of determining one of them very accurately eventually leads to uncertainty about the value of the other. In order to see why this happens, Heisenberg considered some examples of the methods used to measure position and momentum.

One way to measure the position of an object is to shine light on the object, and then focus the light with a lens to obtain an enlarged image of the object. The position of the object relative to some fixed point can then be determined from the image. In order to do this accurately, a sharp image is required; that is, we must be able to distinguish nearby points in the image from each other. It has been known since the nineteenth century that for a microscope, or any other optical instrument, the sharpness of the image depends on the color of the light being used. For light of any specific color, no improvement in the design of the instrument can ever sharpen the image beyond a certain minimum, so that two points in the object closer than some minimum distance cannot be distinguished. Therefore, light of that color cannot be used to determine the position of the object to a degree better than some minimum value.

To understand why this occurs, we must realize the wave properties of a light beam which were discovered in the early nineteenth century. Essentially, a beam of light of some definite color is not homogeneous in its properties. Instead, the intensity of light varies from point to point over very short distances along the beam. The intensity repeats its value periodically over a distance called the wavelength of the light (Fig. 10), which depends only on the color. These wavelengths can be measured in several ways. One is by reflecting light of a definite color from a metal strip upon which a series of fine grooves have been cut at regular intervals. Such a strip is called a diffraction grating. According to a mathematical analysis of the diffraction grating, the light will be reflected at certain angles that depend on the spacing between the grooves and on the wavelength. Therefore, the wavelength can be

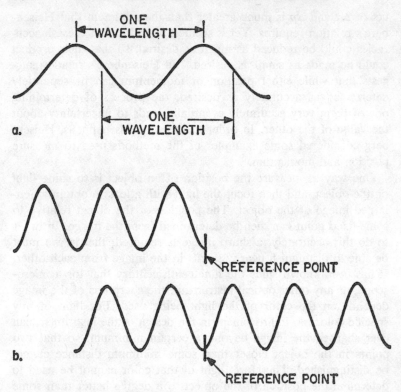

FIGURE 10. Wavelength and frequency of waves. The wavelength of a light wave, or any other wave, is the distance between two successive peaks of the wave, or between any two successive points of equal intensity, as shown in a. If the wave is moving, the frequency is defined as the number of wave cycles that pass any point in space in one second of time. In b. of the figure, a wave is shown at two different times, approximately 10^{-10} sec apart. In that time, the wave has moved about 3 cm, and exactly 3 full cycles have moved past the reference point, so the frequency of the wave is 3 divided by 10^{-10} sec, or 3×10^{10} cycles/sec. The product of wavelength and frequency gives the speed of the wave, in this case 3×10^{10} cm/sec, the speed of light.

determined by measuring the way the light is reflected. For the light we call visible, the wavelength ranges between 4×10^{-5} and 7×10^{-5} cm, going from violet light to red light. According to the wave theory of light, and also from experience with actual instruments, the minimum distance that can be resolved with a microscope using light of any definite wavelength is approximately equal to the wavelength. Therefore, a microscope or other optical instrument using visible light cannot be used to measure distances, or see objects, smaller than about 4×10^{-5} cm. This distance is of course much larger than the size of an atom, and is even larger than that of the living creatures called viruses. Therefore, some other method must be used even to determine whether a particular electron is in an atom or not.

We have seen that there exist other forms of radiation similar to visible light but with much smaller wavelengths. One such form is the X rays, whose wavelengths are around 10^{-8} cm. Another such form is the gamma rays, whose wavelengths can be as small as 10^{-10} cm. While it is not quite feasible to make a microscope using such radiation, essentially because lenses to focus them are unavailable, the radiation can be used to measure distances or produce images by other, indirect methods. For instance, a kind of diffraction grating like that described above for visible light can be made using the regular arrangement of atoms in a crystal in place of the ruled grooves to reflect the X rays. By measuring the reflection pattern of X rays of known wavelength, information can be obtained about the spacing between atoms in the crystal, which is typically about 10^{-8} cm, or about the typical wavelength of the X rays. By this technique of X-ray diffraction, Maurice Wilkins and Rosalind Franklin were able to get a detailed description of the arrangement of atoms in the DNA molecule, providing the necessary information for James Watson and Francis Crick's double helix model of DNA.

It therefore seems possible, by using radiation of smaller and smaller wavelength, to measure the position of objects as accurately as desired, at least over the scale of distances occurring in atoms. Suppose we imagine measuring accurately the position of an electron in an atom several times in rapid succession. We might then expect to be able to find both the orbit of the electron and the electron's velocity through use of the relation that the velocity is the distance moved divided by the elapsed time. This method would be analogous to the way astronomers determine the orbit of a planet by successive measurements of its position in the sky. But an important difference occurs between the two cases. The light reflected from a planet has a negligible influence on the planet's motion because of the huge mass of the planet. We cannot so readily assume that an X-ray beam reflected by an electron will not affect the position or the motion of the electron. On the contrary, we have reason to think that such an effect indeed occurs.

One reason for this asumption is an observation first made by Arthur Compton. It was found that when X rays hit electrons in matter, the X rays often scatter (Fig. 11) through an angle in such a way that they change their wavelength, and at the same time electrons are ejected from the matter at high velocity. Compton showed that this could be accounted for quantitatively by assuming that X rays of a definite wavelength contain particles of a definite momentum equal to Planck's constant divided by the wavelength. When one of these particles of light, called photons, strikes

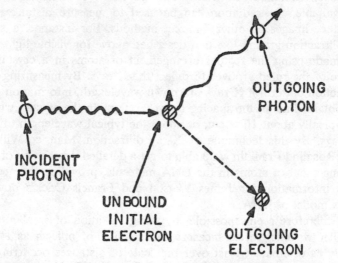

FIGURE 11. Compton effect. In the Compton effect, a photon, here symbolized as a circle with an arrow and an associated wave, collides with an electron, symbolized by a circle with arrow and no wave. The photon loses some of its energy, and so decreases its frequency, and emerges with a longer wavelength, as indicated. The electron gains the energy, and is ejected. The spins of electron and photon, represented by the arrows passing through the circles, also may change.

an electron, it loses some of its momentum, which is transferred to the electron in much the same way as occurs in a collision of two billiard balls, one originally at rest, the other in motion. The scattered photon, having less momentum, correspondingly has a larger wavelength. Subsequent measurements have shown that other forms of radiation such as light also contains photons with a momentum related to the wavelength.

A particle that carries momentum should also carry some energy. Indeed, since the ejected electron carries away some energy, the conservation of energy would suggest that the scattered photon, with its larger wavelength, also has less energy. Compton's analysis implies that the energy of the photon is equal to its momentum, multiplied by the speed of light. The relation between the energy and momentum of a photon is a consequence of the theory of relativity which I will discuss in Chap. V. It follows that the energy of a photon is equal to Planck's constant times the speed of light divided by the wavelength of the X ray. The speed of light divided by the wavelength of a specific type of radiation is called the *frequency* of that radiation (see Fig. 10). The frequency of a wave is the number of wave crests that pass a given point in a second. The shorter the wavelength becomes, the higher the frequency. For visible light, the frequency is about 10^{15} cycles per second (c/sec); for X rays it is about 10^{18} c/sec. The energy of a photon is therefore equal to Planck's constant multiplied by the frequency of the wave with which the photon is associated. This relationship was actually proposed by Planck at the beginning of the twentieth century in order to explain the distribution of wavelengths in the light emitted by glowing objects.

The relationship between momentum and wavelength for photons is one aspect of a more general relationship that also occurs for other particles such as electrons. This more general relationship is essentially the de Broglie relation for electrons and other particles we have previously stated. Compton's discovery actually came before de Broglie's generalization; the wave properties for light had been known previously, and the particles were the new phenomenon, whereas for electrons, the opposite was the case. Perhaps the most remarkable aspect of the discovery of photons is the recognition that light beams actually have a "granular" structure, and that, just as for matter, what appears to be a continuous flow is composed of large numbers of particles. If a beam of visible light has a power of 100 watts (W) and an area of 10 cm^2, it will contain about 10^9 photons/cm^3. Therefore a light beam is a much more rarefied object than even a dilute gas.

We have seen that in order to measure the position of an object very accurately by reflecting radiation from the object, we must use radiation of very short wavelengths. This radiation will contain photons with a large momentum, and in the process of reflection some of this momentum will be transferred to the object. In other words, a precise measurement of position will produce a large change in the momentum of the object being measured. The change in momentum of the object whose position is being measured will be approximately the momentum carried by the photon, and will be in some unknown direction. If we multiply this induced uncertainty in the momentum by the uncertainty in the measured position, which is proportional to the wavelength of the radiation, then, because of the relation between the momentum and wavelength of a photon, the result is a number equal approximately to ½ $\hbar$ as required by Heisenberg's relation. If we try to measure the position twice in rapid succession in order to determine the velocity, the change in momentum that is induced will lead to a change in velocity as well, so that the value obtained for the velocity by this method will have little relation to the velocity before the measurement occurred, or after it is over. Therefore, this method does not seem feasible for determining the position and velocity of an object at the same time. We note, however, that since it is the momentum of the object that is changed by its reflection of the photon, and the change in velocity is related to the change in momentum through dividing by the object's mass, a very heavy object, such as a macroscopic body, will have little change in its velocity through measurement of its position, while a small object, such as an electron, will undergo a much greater change in velocity. All of this is quite consistent with the requirements of Heisenberg's relation.

One method that fails to work is no proof of a general rule, and we should explore other possibilities. Actually, the discussion just given implies that velocities of small objects cannot really be measured at all by separated position measurements. However, there are other ways to measure velocity. For example, there is the method used by police to determine if cars are speeding along a highway and by flight controllers to measure the speed of an airplane. This involves reflecting radiation of known frequency from the object, and determining the change in frequency of the radiation that is reflected back at the sender. According to a result of relativity theory (discussed in Chap. V), this change in frequency is proportional to the velocity of the object toward or away from the sender. The phenomenon involved is known as the *Doppler effect*. The result of such reflection of light can be obtained in the same way as in the analysis of the Compton effect, except that the object reflecting the light is taken to be moving, rather than at rest. In accordance with our previous discussion, we should use radiation of long wavelengths in this method of measuring the velocity, because the photons in such radiation will have small momentum and therefore not disturb the velocity of the object being measured. Furthermore, since the velocity measurement depends on determining the change in wavelength of the light reflected, this light must have a wavelength that is known much more accurately than the change that occurs on reflection. But a light wave with a very precisely defined wavelength must extend over all of space, as we mentioned in our discussion of the electron wave packet. When such a wave is reflected, it can happen anywhere in space, or, correspondingly, at any time. This means that after the reflection, we know the velocity of the electron, but do not know its position when it has this velocity. The uncertainty in the position is just the uncertainty in the time of reflection multiplied by the velocity. A simple calculation shows that the product of this uncertainty in position, and the uncertainty in momentum after reflection, due to the momentum transferred from the photons to the particle, is always greater than $\frac{1}{2}\hbar$, and thus Heisenberg's relation also holds for this method of measurement. Actually, the content of this relation is that the methods we can use to measure positions necessarily alter the momentum of an object in an uncontrollable way, and, conversely, Heisenberg's relation gives a quantitative measure of how much information about one of these qualities we must lose upon trying to improve our knowledge of the other. The effect is unimportant on a microscopic scale, but is essential for atomic or subatomic phenomena.

Our discussion of the rationale for the Heisenberg relation has thus far been in terms of using light or other radiation to measure position or momentum. Other methods for making such measurements which might escape the restrictions that are connected with the use of light exist, and are commonly used by scientists. For instance, beams of electrons themselves are used to make images of very small objects such as viruses in the device known as the electron microscope. In this device, electrons pass through an object being viewed, are focussed by magnetic forces in a way similar to the focussing of light by glass lenses, and images are formed by the focussed electrons in the same way as the focussed light rays form images in an ordinary microscope. It is possible to image very small objects with electron microscopes, but limits exist on the accuracy with which this can be done. These limitations arise from the de Broglie waves associated with electrons, and the same arguments that apply for light also imply that electrons of some definite wavelength cannot form good images of objects smaller than this wavelength. This means that in order to image very small objects with an electron microscope, we have to use electrons of very short wavelength, and therefore with large momentum and energy. Such electrons tend to transfer large amounts of energy and momentum to the atoms in the object being imaged as they pass through the object. Therefore, just as in the use of ordinary microscopes, an accurate measurement of the positions of objects with an electron microscope will produce an uncertain change in the object's momentum. Because the relation between wavelength and momentum is the same for electrons as for photons, the product of the uncertainties in position and momentum will again be approximately that required by Heisenberg's relation.

It is of course not possible to prove Heisenberg's relation by considering all possible means of measurement and their associated uncertainties. Rather, physicists take the view that this relation is an essential element of the laws of nature, and that any measuring instrument we may develop will necessarily have properties conforming to it. If the wave properties of electrons had not been known before, the analysis of measurements with electron microscopes would have shown that either such properties must exist, or that Heisenberg's relations, and the description of atoms obtained with them, are false. Another way of putting this is that the rules of quantum mechanics are mutually consistent in such a way that they are either true for all kinds of objects that can influence each other, or they are generally false. These rules cannot be true for some objects and not true for others without leading to contradictory results.

The wider implications of Heisenberg's relation have been the subject of much analysis and discussion by both philosophers and physicists. One implication, about which there is general agreement, is that the description of the world given by quantum mechanics is not deterministic. That is, for an atom, or any subatomic system, there exist measurements that will give definite outcomes when they are made, but whose results cannot be predicted in advance from any information we can have about the system. An example is the scattering of an electron by an atom. The electron directed toward the atom along some line will emerge along another line making some angle with the incident line. This angle cannot be predicted for a given electron scattering, although the probability of scattering through various angles can be predicted and tested by scattering many electrons. This is to be contrasted with the corresponding situation in Newtonian physics according to which the angle at which a given electron is scattered is fixed by the velocity of the incident electron and the distance of its closest approach to the nucleus. Obviously, it is the Heisenberg relation that keeps us from learning these two quantities with enough accuracy to make the prediction. There *are* circumstances in which exact predictions can be made in quantum mechanics. One such is that the momentum of an isolated system, if known to have some value at one time, will definitely have the same value at a later time, if measured. However, this is an exceptional case, and in general only probability predictions can be made.

Quantum mechanics therefore implies a restriction on our capability of predicting the future, which was not a part of Newtonian physics. In the nineteenth century, Pierre Laplace summarized the determinism of Newtonian physics with the image of a demon who was both a supermeasurer and a supercalculator. This demon measured the positions and velocities of all the objects which composed the Universe. It then used the Newtonian laws of motion to calculate how all these objects would move in the future, since according to these laws the positions and velocities at one time determine the positions and velocities at all future times. The demon could, therefore, predict the entire future of the Universe. Of course, there are insuperable practical difficulties in doing this even for the molecules in a small volume of gas, but some indication of Laplace's vision can be gotten from the fact the astronomers can predict the future positions of all the planets for thousands of years on the basis of present observations.

However, no quantum-mechanical equivalent of Laplace's demon can exist, if only because the information that Laplace required his creature to have is unavailable because of Heisenberg's relation. According to quantum mechanics, there are things which we can know when they happen, and can predict the probability of happening, but not anticipate with certainty beforehand. In this sense quantum mechanics gives us a more restricted understanding of the atomic world than we might like to have. We can perhaps take some comfort in recognizing that it gives us many insights into the behavior of these objects. We may be distressed, but should not be surprised, that what we can learn about the behavior of subatomic particles is less than what we know about macroscopic objects. In a slightly different vein, we should not be surprised that the notion of exact prediction of the future behavior of a system, abstracted from the motions of large bodies under simple forces, should not be completely extendable to another situation involving completely different orders of magnitude. Rather, it is surprising that most of the notions of Newtonian mechanics are applicable in the atomic domain. It is one of the miracles of human creativity that given the immense disparity of sizes involved and the differences in behavior between atoms and ordinary bodies, our minds have been able to frame the right concepts to explain the new features of atomic phenomena. Some further examples of these explanations will be given in Chap. IV.

IV

Light and Atoms

A question of some importance is how physicists can measure the internal properties of atoms. A doubt that such measurements could be accomplished made some nineteenth century scientists skeptical about the very existence of atoms. Probably the most important tool used to make such measurements in the interaction of the atom with light of various frequencies. This interaction can take several forms (Fig. 12). Light may be spontaneously emitted from an atom when an electron changes its state. Alternatively, a light beam may be directed toward an atom from outside, and either scatter from the atom or be absorbed, causing some change in the atomic state. By combining various types of these measurements, physicists have learned a good deal about the internal atomic states, and have often been able to verify the predictions of quantum mechanics, as well as provide new data for theoretical analysis.

We have seen that a light beam of a definite frequency is made up of a large number of particles of definite energy and momentum, the photons. For a beam of visible light, the photon energies are comparable to the energy of the outer electrons in the atom. Therefore, the discussion of the interaction of visible light with atoms should be in terms of photons rather than in terms of unquantized light waves. This is equally true for higher frequency light, such as ultraviolet or X rays. We will first consider the simplest process, emission of light by the atom.

When an atom is in its ground state, in which all the electrons

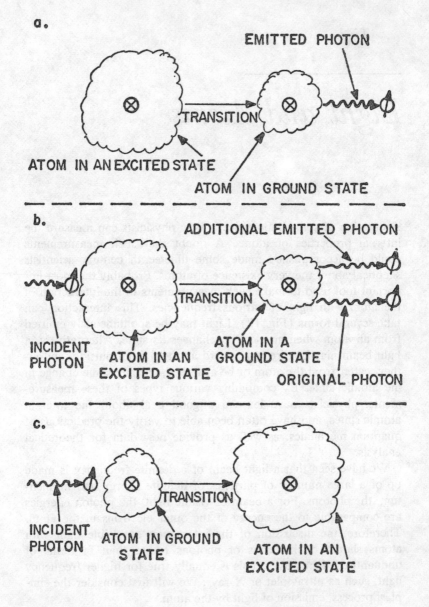

a.

EMITTED PHOTON

TRANSITION

ATOM IN AN EXCITED STATE

ATOM IN GROUND STATE

b.

ADDITIONAL EMITTED PHOTON

TRANSITION

INCIDENT PHOTON

ATOM IN AN EXCITED STATE

ATOM IN A GROUND STATE

ORIGINAL PHOTON

c.

TRANSITION

INCIDENT PHOTON

ATOM IN GROUND STATE

ATOM IN AN EXCITED STATE

occupy the lowest-energy orbits allowed them by the exclusion principle, the atom cannot emit any photons, because the photon would carry away a positive amount of energy, and this would require the atom's energy to decrease in order to conserve the total energy. If the atom is already in its lowest-energy state, it is clearly impossible for this to happen, and thus no photons are emitted from the atomic ground state. Sometimes, however, one or more electrons are raised to an excited orbit whose energy is higher than that of the orbit in the ground state. This might happen through a collision with another atom, with an electron from outside the atom, or from other disturbances. For atoms in a typical solid body at room temperatures, such disturbances occur about once in 10^{-12} sec, or very often on an ordinary time scale. On the other hand, the time for one "revolution" of the electron in its orbit is about 10^{-16} sec, so that the disturbances occur relatively about as rarely as large comets appear in the solar system.

When an electron is in an excited state, it can emit a photon and drop down to any unoccupied lower-energy state. The length of time this takes to happen depends on the two states involved, but usually is 10^{-8} sec or less. A change in electron states of this type is called a *transition*. Such transitions will occur whenever lower-energy states are available for the electron to drop into. Consequently, the excited state is unstable, and the electron remains there only until a transition occurs. Only the ground state is stable in the sense that the atom will remain there indefinitely in the absence of external forces. When an electron changes states in

FIGURE 12. Three forms of interaction of light with atoms. a. An atom in an excited state can emit a photon, while making a transition to a lower state, such as the ground state. This can happen without any outside stimulus to the atom, in which case it is known as spontaneous emission. b. If a photon whose energy is equal to the difference between that of an atom in an excited state and the atom in a lower state hits the atom in an excited state, the atom may emit another photon of the same energy, so that two such photons are present afterward, and make a transition to the lower state. This is known as stimulated emission. c. If an atom in its ground state is hit by a photon whose energy is the difference between the energy of the atom in some excited state and in the ground state, the atom may make a transition to an excited state, while the photon disappears. This process is known as absorption.

this way, the photon that is emitted will have an energy equal to the difference of energy of the two atomic states. Since an atom of a given element has only certain possible energies, corresponding to the allowed electron states, this atom can only emit photons with certain energies, and hence certain discrete frequencies. Historically, the fact that energy levels in the atom are quantized was inferred by Bohr on the basis of the occurrence of discrete frequencies of light emitted by the atom.

The photons emitted by atoms in an excited state in returning to the ground state are what is actually observed in the so-called emission spectrum of the substance formed by these atoms. For example, if the substance is boiled into a gas, and an electric current discharged through the gas, some number of the atoms will be raised to excited states, through collisions with the moving electrons that constitute an electric current, and then emit light as they decay to their ground state. This light is characteristic of each type of atom, as indicated by the different color of the glow emitted by neon lights, mercury-vapor lights, etc. The exact pattern of the light can be studied with a device called the spectroscope, which uses a prism to separate the light into its many component colors, each of which corresponds to a wave of a definite wavelength. The particular pattern of colors emitted by a specific element (or compound) is unique, and rather independent of how the atoms are excited. It is this pattern of colors that we call the emission spectrum of the element.

When seen through a spectroscope, the emission spectrum of an element consists of a set of separate thin bands, each centered in a definite wavelength, rather than a continuous spread of colors such as the rainbow or the spectrum of an incandescent light bulb. Different atoms will generally have different emission spectra because their energy levels are different. Once the spectrum of a particular atom has been mapped out, it can be used to identify the atom in novel circumstances. For example, it is possible to know that hydrogen atoms are found in distant galaxies by measuring the frequencies of light reaching Earth from these galaxies and identifying some of these frequencies with those in the hydrogen spectrum. Occasionally, a frequency is detected that does not fit in the spectrum of any known atom or molecule. In one case, this was used to identify the previously unknown element helium in the Sun's atmosphere. In other cases, the unexplained frequency corresponds to the emission of light by an ion of some element rather than an atom. The ion has different energy levels than the neutral atom, and hence has a different emission spectrum.

To study the spectra of atoms, it is usually convenient to raise the temperature of the material containing the atoms. This has the effect of increasing the rate at which collisions occur since the atoms are moving faster. This in turn increases the number of atoms that have electrons in the excited states and therefore the rate at which the photons are emitted. The motion also has slight effects on the energies of the photons emitted, such as the Doppler effect, to be discussed later. Spectra of atoms are best measured from atoms in a gas, or even more rarefied forms. This is because when there are many atoms present in a small region, as in a solid, the atoms can affect one another, as well as the light they emit, in such a way that the light emerging from a solid body bears little resemblance to that which the atoms composing the body would emit in isolation. Of course it is interesting to study the light emitted by solid bodies also—and that is another phenomenon to be explained—but it only indirectly relates to the properties of the atom. In a gas, where the atoms are far apart, the light emitted by each atom is unaffected by the other atoms so that the spectrum observed is essentially that of isolated atoms.

By measuring the energies of the various photons emitted by an atom, we learn about the differences of energy between the various electron states in the atom. In order to learn the actual energy of the states, rather than the differences, it is necessary to measure the energy of any one state by some other method. I will describe one such method below. The electron energies that are inferred from the photon energies by this procedure may be compared with theoretical calculations of the energies based on Schrödinger's equation. These comparisons generally result in good agreement. In the special case of the hydrogen atom, which is sufficiently simple to allow very accurate calculations, the comparison between theory and experiment has reached the fantastic precision of one part in 10^{11}, and the agreement persists. Obviously, the quantum theory of the hydrogen atom is a great success in this respect.

The photons emitted in a transition have both spin and orbital angular momentum as well as energy associated with them. Since total angular momentum is conserved just as energy is conserved, the atom's angular momentum must change in order to balance that carried away by the photon. In most atomic transitions, the photon emitted carries away one unit of total angular momentum, corresponding to its spin of one unit. Such transitions are called allowed transitions. Not all pairs of electron states can be linked by an allowed transition because they may not differ by one unit of angular momentum. Transitions between such states can usually still occur, by having the photon carry off several units of angular momentum, giving what is known as a forbidden transition. This name is unfortunate, as the transition is clearly not forbidden if it occurs. However, it does take longer to occur, and, until recently, forbidden atomic decay transitions had only been detected in various un-earthlike conditions. The reason for this is that on Earth an excited atom may de-excite in the same way as it was excited—by a collision with another atom. In an ordinary gas this is unlikely to occur in the short time that an allowed transition takes, but can easily happen before a forbidden transition can occur. However, in the solar corona, or in interstellar space, the density of atoms is very low, and the collision probability is greatly reduced compared to that on Earth, so there is time for forbidden transitions to occur. From the average length of time it takes for a transition to occur, it is possible to determine the angular momentum carried off by the photon. This then determines the change in angular momentum of the atom, which is further information about the angular momentum of each state involved in the transition. The angular momentum of the atomic ground state can usually be measured by other methods, and the angular momentum of the excited states inferred through the information obtained from transitions.

The transitions we have been discussing are often called spontaneous transitions, because they occur even if the atom is isolated after the electron is excited to a higher-energy state. In other words, the emission of the photon does not depend on anything outside the atom. Rather, it is as if there is an internal mechanism in the atom that makes it emit the photon. It is obviously of interest to know whatever we can about how this emission takes place, and a number of things can be said on the basis of the quantum theory. But certain other seemingly plausible questions do not have answers according to this theory, just as some questions about the motion of the electrons in the atom cannot be answered. For example, we might want to know how long an electron that is excited to some higher-energy state will remain there before emitting a photon and returning to the original state. The quantum theory implies, and experiment suggests, that this length of time is unpredictable for an individual atom. However, if the process is repeated for many atoms, always exciting the electron to the same state, there will be a definite pattern to the times required for the transition to occur. Provided that the number of atoms is large enough that the laws of probability apply, it will be found that about half of them will emit photons in a definite time, characteristic of the transition. Half the remaining excited atoms will again emit photons in this time, etc. This behavior is obviously analogous to the decay of radioactive atoms, and the time in which half the atoms will emit a photon is therefore called the half-life of the transition. It should be recognized that an individual atom may take a much longer or a much shorter time to emit the photon than the half-life of the transition, but the probability of either of these is small. Given any large number of atoms in an excited state, about 99% of them will have made transitions to a lower state within a period of seven half-lives. If all of the 6×10^{23} atoms in a gram of hydrogen gas could be put independently in the same excited state at the same time, it would take about eighty half-lives until they had all made transitions.

The process by which the electron makes the transition from one state to another and the photon appears cannot be understood in terms of Newtonian mechanics any more than Heisenberg's relation can. The electron does not move continuously from one state to another, since any measurement of the electron's energy must give either the value in one state or in the other, not some value between. Furthermore, a measurement to detect whether the photon has been emitted must yield only an answer of yes or no, since it is no more possible to find half a photon than to find half an electron. Because of this, we must imagine the emission of the photon, together with the change of state, as a single event, occurring discontinuously rather than gradually. Indeed, this is the way we should think of all changes according to quantum mechanics. The change in the actual physical situation, such as the energy state or the type of nucleus in a radioactive decay, occurs suddenly and at random. The thing that changes continuously and according to definite laws is the probability that the system is in one or another of its possible configurations; for example, an electron in an excited state with no photon present, or an electron in the lower state with a photon present. The half-life of a transition refers to the time that it takes for the first of these probabilities to reach ½, not to the time the actual change requires. In order to predict the time of occurrence of an individual transition, it is natural to imagine that we could make measurements on the atom just before the transition occurred that would reveal whether or not the atom was "ready" to emit the photon. However, such measurements would grossly disturb the motion of the electrons in the atom, so that its subsequent behavior would have no necessary relationship to that of an undisturbed atom. The randomness in the time at which an individual atom makes a transition is therefore analogous to the randomness of the result of measuring the position of an electron in an atom that follows from Heisenberg's relation.

The unpredictability of the time at which an atom will make a transition is therefore another example of how quantum mechanics limits the extent to which we can make accurate predictions about the future. It is not that this time is unknowable, since if we wait until the transition has occurred we can easily determine when it happened, for instance, by measuring how far the photon has moved when it is detected. It is rather that we have no way of knowing the time in advance. This and other features of quantum mechanics led Einstein to believe that it was an incomplete theory, and would eventually be replaced by another one in which such things as the time at which a particular transition will occur could be predicted exactly. Most other physicists doubt this, and remain convinced that randomness in the occurrence of individual atomic or subatomic events is a fundamental feature of the world, rather than an artifact of human ignorance. None of the developments of physics in the fifty years since quantum mechanics was invented has given reason to think otherwise.

The Absorption of Photons by Atoms

Another way in which light can be used to get information about the structure of atoms is to direct a light beam at a group of atoms and measure what happens to this light. We first imagine that the atoms are almost all in their ground state, as this is the usual situation. One possibility is that a photon from the beam is absorbed by an atom. This can happen in several ways. If the photon energy is exactly the difference between the energy of the atom in its ground state and the energy the atom would have in some excited state, the photon can be absorbed, leaving the atom in this excited state. This process is known as resonance absorption and the photon energy is called a resonance energy. Actually since the excited state is unstable, the atom may later make a transition back to a lower state, re-emitting a photon in doing so. If the latter transition is to the ground state, this photon will have the same energy as the photon that was originally absorbed. However, there may be other excited states with energies between the ground state and the original excited state, and so the atom may make a transition to one of these, emitting a photon with lower energy than the one it absorbed. As a further alternative, if the atom is in a solid or a liquid, the atom in the excited state may lose energy by collision with another atom, and not emit a photon at all. In this case, the energy of the photon that was absorbed ends up converted into the heat energy of random motion of the atoms in the body.

This type of resonant absorption of light gives the same type of information about the atom as does the emission of light. In fact, the two processes are related to one another just by reversing the time order in which events occur. However, there are circumstances in which absorption is a more convenient way of getting information than emission. For example, suppose that there is some excited electron state for which the transition to the ground state is a forbidden transition. For reasons described before, this transition would be hard to measure by the emission of light. On the other hand, it is possible to expose the atoms in the ground state to light containing photons of exactly the right energy to be absorbed, thus placing an electron in the excited state of interest. The fact that the transition is forbidden will decrease the probability of this happening, but this can be overcome by increasing the intensity of the light beam. Many forbidden transitions have been studied and the properties of the states determined by this absorption technique.

Suppose that the photon energy of the light hitting the atom is not equal to the energy difference between two states of the atom. Then the photon cannot be absorbed by the atom unless its energy is so high that an electron absorbing it can be removed from the atom altogether. This will first be possible for one of the outer electrons, usually at a photon energy of several eV. This is the so-called ionization threshold of the atom and occurs at the same energy that we have described in the Franck–Hertz experiment with electrons. The process by which an electron is removed from the atom by absorbing a photon is known as the photoelectric effect, or photoionization. Photoionization can occur with photons of any energy above the ionization threshold of the atom. Any energy above the minimum needed for photoionization ends up as kinetic energy of the electron removed from the atom. Because of the relation between photon energy and light frequency, this electron kinetic energy will be simply related to the frequency of the light. Accurate measurements of this relation between electron energy and light frequency by Robert Millikan helped establish the truth of the relation between frequency and photon energy.

Actually, there are a series of ionization thresholds at higher and higher photon energies corresponding to removal from the atom of electrons in orbits of lower and lower energies. In order to determine the energy corresponding to these thresholds, we can imagine exposing atoms to a series of photons of higher and higher energy. As the photon energy reaches each new threshold, the probability that the photon is absorbed by the atom increases drastically. When the photon energy increases beyond the threshold energy, the absorption probability decreases slowly, until a new threshold is reached. The final threshold corresponds to the removal of the electrons in the lowest-energy states, and above this energy the probability of absorption by photoionization decreases steadily. By measuring the photon energies at which the absorption increases suddenly, the energies of the electrons in the various orbits can be determined, as the conservation of energy implies that when the photon is absorbed, its energy all goes to the electron that is ejected. In this way, the energy of electrons in each of the orbits of the atomic ground state can be measured. Combining this with the measurements of energy differences from photon emission or resonant absorption, we can determine the energies of many possible electron orbits of any atom.

When this is done, it is found that the binding energy of the outermost electron orbits in the atomic ground state are usually a few electron volts in almost all atoms. The inner orbits, however, have much larger binding energies, which tend to increase as the electron numbers increase. The binding energy of the two innermost electrons, known as the K-shell electrons, increases approximately as the square of the number of electrons (or equivalently, of the nuclear charge). This binding energy may be several hundred thousand eV in the heaviest known atoms. The reason for this behavior is that the outer electrons are attracted by a force that is the combination of the attraction of the nucleus, and the repulsion of the inner electron shells. The result of this is that the potential energy of these electrons has a value that depends very little on the nuclear charge, because all but a few units of the nuclear charge are effectively cancelled by the inner electron shells. This is called screening the nuclear charge. Since the binding energy is approximately proportional to the potential energy, there will be no tendency for the binding energy of the outer electrons to increase as the nuclear charge increases. Further, since the binding energy is simply related to the average distance of the electron from the nucleus, this implies that the outer electrons of all atoms are approximately the same distance from the nucleus, about 10^{-8} cm. This helps explain the fact that in most solid bodies the number of atoms per unit volume is about the same, since the average spacing of the atoms is determined by the radius of the outer electron orbits. Of course, outside the outermost electron orbits the screening is complete, since the electron charge exactly compensates that of the nucleus. Hence, to an external object, the atom acts like a neutral body. Nevertheless, since the positive and negative charges are not quite at the same point, there are small residual electric forces outside the atom which are responsible for some types of molecular binding.

On the other hand, the innermost electrons in the most tightly bound shell have the full force of the nucleus acting on them, as there are no electrons nearer the nucleus to screen its charge. For these electrons, the potential energy and the binding energy increase as the nuclear charge increases, and so the average distance from the nucleus decreases as the charge increases. In atoms such as lead, the K-shell electrons are within 10^{-10} cm of the nucleus, which is still far outside the nucleus itself. Because the binding energy of the inner electrons is so large, they are not ordinarily transferred from one atom to another or removed from an atom by collisions at ordinary densities and temperatures. Only the outer electrons are involved in these processes.

The Scattering of Photons by Atoms

If a photon hits an atom in the ground state and the photon's energy is below the lowest ionization threshold and is unequal to any of the resonance energies, the photon cannot be absorbed by the atom. It can be scattered, that is, diverted through some angle, with or without a loss of energy. If the photon loses no energy, the atom must emerge from the encounter with the photon in the ground state. In this case, the scattering is called elastic scattering, or Rayleigh scattering, after Lord Rayleigh, who first studied it. On the other hand, if the incoming photon energy is greater than the lowest resonance energy, the atom can emerge in an excited state and a new photon with a lower energy will emerge. This process is called Raman scattering after Chandrasekhara Raman, or inelastic scattering. Actually, scattering can occur for photons of any energy, even above the ionization threshold. But usually for photons, if both scattering and absorption can occur, the probability of scattering is much less than that for absorption and the scattering is difficult to detect. These probabilities can be made precise through the concept of a *cross section* for the process. A cross section is a quantity with the dimensions of an area, measuring the effective size with which a specific target causes a beam to undergo some process. In the simplest case of a target which is a large opaque disk, absorbing all the light hitting it, the cross section would be just the geometric area of the disk. If the disk were somewhat transparent, only some of the light would be absorbed, and the cross section would be less than the area. This is approximately the situation for absorption of visible light by the atom, in which case the cross section is usually around 10^{-24} cm^2 whereas the geometrical area is around 10^{-15} cm^2. However, the absorption cross section can be larger; in the case of resonance absorption, as large as 10^{-9} cm^2. Such supergeometric cross sections are possible only because of the wave properties of photons, or of other particles, through the effect called diffraction.

The scattering of light by an atom has a cross section that rises rapidly as the photon energy increases. For red light, the cross section is around 10^{-28} cm^2, while for blue light it is approximately ten times larger. These results were used by Rayleigh to explain why the Sun's light appears redder when the Sun is near the horizon, and why the light of the sky appears blue. The reason is that although the Sun emits similar amounts of red and blue light, the red light is scattered less in passing through the atmosphere than is the blue. Therefore, the direct sunlight contains more red than blue, while the light of the sky, which is all scattered sunlight, contains more blue than red.

Rayleigh and Raman scattering are more complicated to describe theoretically than the absorption of photons. One way of conceptualizing them is to recognize they take place in two steps which can occur in either order (Fig. 13). In one step, the incident photon is absorbed by the atom, while an electron makes a transition. In the other step, the atom emits the outgoing photon, and an electron makes another transition. The combined result of the two electron transitions must take the atom from the ground state to whatever state in which it emerges, but there is no restriction on the energy of either transition. The two-step character of these scatterings suggests that the cross section for scattering should be less than that for absorption, if both are possible, since the probability of two steps following one another is generally less than the probability of either step.

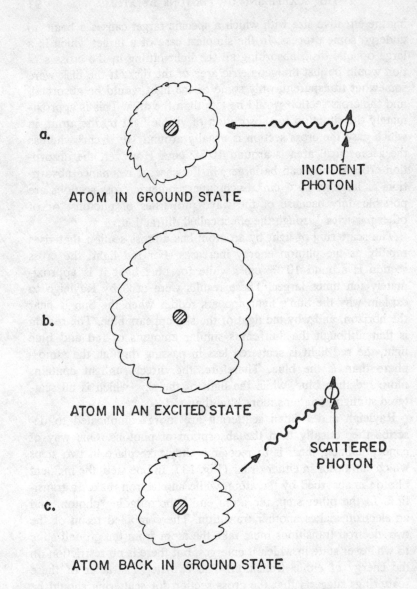

a.

ATOM IN GROUND STATE

INCIDENT
PHOTON

b.

ATOM IN AN EXCITED STATE

SCATTERED
PHOTON

c.

ATOM BACK IN GROUND STATE

The description just given of scattering seems to contradict either the conservation of energy or the idea that the energy levels of atoms are quantized. If the incident photon energy is not equal to a resonance energy and is not above the ionization threshold, how can it be absorbed as one step in the scattering? And how could the atom emit the outgoing photon before absorbing the incoming photon, if the atom starts in its ground state? The answer is that scattering is a process that occurs in a very short period of time—less than 10^{-17} sec. Therefore, any transitions made by the atom would be within this time period. An analysis by Heisenberg, related to the one that led to Heisenberg's relation for position and momentum, implies that if a system like an atom is in some condition for a limited period of time Δt, then its energy while in this condition is not determined completely, but is instead uncertain by an amount ΔE, such that ΔE times Δt is greater than $\frac{1}{2} \hbar$. This relationship between ΔE and Δt is sometimes known as Heisenberg's relation for energy and time. From it, we can see that during the time of the transition of the atom to the excited state, the energy of the atom is uncertain by an amount that is approximately several hundred eV, and there is no problem with energy conservation during the process as this uncertainty is much greater than the energy-level differences of the atom. If we were to try to measure the energy of the atom during the short time interval of the transition, we would alter the scattering process considerably. For example, if the atom were found to be in its ground state in the middle of the transition, there would be no scattering

FIGURE 13. Scattering of light as a two-step process. The scattering of light by an atom occurs as a two-step process if the energy of the photon is not equal to the energy difference between two states of the atom. In the first step, the photon is absorbed by the atom, which makes a transition to an excited state, as in b. In the second step, the atom emits a photon with the same energy, but different direction than the first photon, and returns to the ground state. Since the energy of the excited state is not equal to the original energy of atom and photon, the transitions from a. to b. and from b. to c. are virtual transitions, and the excited state persists only for the short time allowed by Heisenberg's relation.

of the photon for that particular case. On the other hand, if one measures the energies of atom and photons long after the transition occurs, the quantity Δt can be made very large, as it now refers to the final atomic state which persists indefinitely. Therefore, ΔE can be small, and energy conservation will be found to hold accurately. This illustrates the importance of precisely specifying what measurements are and are not made on a system in order to be able to describe its behavior according to quantum mechanics.

The transitions to intermediate states that do not conserve energy over the very short time period of the over-all scattering process are known as *virtual transitions,* as opposed to the real transitions that satisfy the conservation of energy. Virtual transitions play an essential role in the quantum-mechanical description of how a situation evolves with time. In many cases, as in the photon scattering processes, the over-all change from initial to final state occurs mainly because of some virtual transitions that bridge the states through intermediate states. In other cases, the over-all process can happen both directly *and* through virtual transitions. In order to calculate the over-all probability for the scattering, it is necessary to assume that the process actually may occur via virtual transitions to any of the intermediate states and the results combine to give the over-all probability. In fact, one must even include longer chains of virtual transitions, in which after one intermediate state is reached, there is a transition to a second and then a third such intermediate state, until finally a transition is made to the final state. For the scattering of photons, these more complicated chains are relatively unimportant compared to the two-step process. On the other hand, for scatterings of some other subatomic particles, the multiple chains of virtual transitions can play an important role.

Because scattering involves many possible virtual transitions in an atom, its measurement does not give much information about any one state of the atom. Instead, it generally gives a kind of average behavior of all of the states. For radiation of much lower frequencies than those of visible light, this information is related to the behavior of the atom when an electric force or a magnetic force is applied to the atom from the outside. The reason for this relation is that the quantized nature of the energy and momentum can be disregarded for radiation whose energy is much smaller than any of the differences of energy levels in the atom. The radiation then behaves—as envisioned in Maxwell's theory—as a combination of electric and magnetic forces, and it is not surprising that the atom should respond to that aspect of it.

The Stimulated Emission of Photons

A very interesting aspect of the interaction between light and atoms occurs when the atoms are in excited states. We have seen that ordinarily a very small percentage of the atoms in a body are in excited states for any length of time. However, it is possible to increase this percentage in various ways. For example, if the body is exposed to an intense light beam containing photons with some of the resonant energies of the atoms, many of the atoms in the ground state will be excited. These excited atoms will decay to lower states, and eventually to the ground state. It may happen that there is a state to which the excited state can decay whose energy is between the excited states that were filled and the ground state. Furthermore, transitions from this middle state to lower states may be slow transitions. In this case, those atoms which have reached this state by transitions from higher states, or by other methods, will remain there for relatively long periods. Such states are called *metastable*. Under suitable conditions, it is possible to arrange a situation in which more atoms are in this excited state than are in the ground state for a substantial time interval. This is known as a population inversion. Actually, an inversion can occur between any two states by arranging it such that there are more atoms in the higher-energy state, but usually the term refers to the ground state and some higher state.

Now consider an atom in the metastable excited state. If the atom were isolated, it would eventually emit a photon and drop to the ground state. Suppose however that a photon of exactly the same energy as the difference between the excited state and ground state strikes the atom before this transition occurs. Then a phenomenon known as stimulated emission can occur in which the atom emits another photon of this energy in response to being hit by the first photon, and drops to the ground state. The original photon remains in existence, so that two photons of the same energy are now present. This stimulated emission may occur much more rapidly than the spontaneous emission of the photon. Furthermore, the rate at which stimulated emission occurs increases as the number of photons with the right energy hitting the excited atom increases.

The reason for this rather strange behavior depends on the properties of photons in groups. We have seen that electrons have a tendency to avoid one another. As expressed in the exclusion principle, an electron will not enter a state already occupied by an electron. Photons on the other hand do not behave this way at all. A photon can readily be emitted into a state already containing a photon. Indeed, the results described above suggest that many photons can occupy the same state, and this is so. Furthermore, if two or more photons with the same spin projections are in some region, they will be found near each other more often than if they moved independently, whereas two electrons will be found near each other less often. These results have nothing to do with any forces acting between the photons or the electrons. Instead they are a consequence of the fact that the possible states of systems containing many identical particles automatically impose correlations on the behavior of two or more particles. Photons and other types of particles that tend to cluster with one another are called *bosons,* while electrons and other particles that obey the exclusion principle are called *fermions.* An important discovery in theoretical physics was that a correlation must exist between the intrinsic spin of a particle, and its property as a boson or fermion. Those particles with a spin of $\frac{1}{2}\hbar$, $\frac{3}{2}\hbar$, etc., must be fermions, while those with a spin of 0, $1\hbar$, $2\hbar$, etc., must be bosons. Photons have spin of $1\hbar$, and so follow this rule. The reason for this connection between spin and statistics, as it is known, is complex, but it appears to be a consequence of any theory consistent with quantum mechanics and special relativity.

The occurrence of stimulated emission has been put to use in the remarkable device called the laser. In a simple type of laser, a population inversion is established in a large number of atoms contained in a cavity, which has mirrors at each end. The action begins when a few atoms in the excited state emit photons spontaneously and drop to the ground state. These photons remain trapped in the cavity by being reflected from the mirrors. When a photon hits another atom in the excited state, the atom may emit another such photon by stimulated emission. This photon will in turn produce still more stimulated emissions, by other atoms, and the number of photons will increase continuously. Of course, if the photons instead hit atoms in the ground state, they will be absorbed and lost. A population inversion is necessary in order to make stimulated emission more likely than absorption. The effect of the process taking place in the laser is that the energy that was originally introduced to excite the atoms, producing the population inversion, ends up in the form of many photons each of whose energies is that of the transition from the metastable state to the ground state. The laser is therefore essentially an energy conversion device, not an energy source. These photons can eventually be allowed to escape from the cavity, constituting a very intense beam of light with a precisely defined frequency.

The main feature of the laser is the production of this light with a definite frequency. The beam emitted by the laser can be focussed much more sharply, or transmitted great distances with less attenuation, than light beams produced by other methods. As an illustration, a laser beam has been reflected from a mirror device transported to the Moon, and the return beam detected on the Earth. Strongly focussed laser beams can be used to bring a large amount of energy to a very small area and have been used in eye surgery. Lasers have also been used to carry out very precise measurements of the interaction of light with atoms, to test some of the ideas described above, and to learn about some of the optical properties of specific substances.

All in all, the interaction of light with atoms has probably been the most successful area of twentieth century physics. The basic ideas seem to be completely understood, and can, in certain cases, be used to predict the result of measurements to incredible accuracy. This has become possible through the systematic application of quantum mechanics both to light and to atoms, and has led to a general conviction among physicists of the correctness of this theory, strange as it seemed at its beginning. Armed with this conviction, physicists have used quantum mechanics to analyze various subatomic phenomena, especially those occurring at much higher energies than the few electron volts that are typical of atomic phenomena. We shall see that these subatomic phenomena have also been amenable to analysis by quantum mechanics, especially when the latter is supplemented by the special relativity theory of Einstein.

V

Relativity Theory— A Study of What Is and Is Not in the Eye of the Beholder

Physicists such as Planck and Einstein have said that the aim of physics is to discover the objective characteristics and laws of the world, those that are the same to all viewers, whatever their situation. However, most aspects of the world with which physicists deal are not objective or invariant in this sense. An object viewed from various directions usually appears different. Observers at different points in space, or the same place at different times, generally do not see the same things. The description of the world given by physicists must not only tell about those things that are the same for everyone, it must also tell how those things that are not the same will vary from observer to observer, and must include some way of determining which aspects of the world are objective, and which are not.

The part of physics that deals with these last two questions has come to be known as relativity theory. In its present form, relativity theory was developed in the twentieth century, mainly by Albert Einstein. Earlier physical theories such as Newtonian mechanics had their own associated forms of relativity theory, but those other forms were not completely correct, and have been abandoned in favor of Einstein's. While relativity theory applies to all physical phenomena, not just those involving atoms and their constituents, we shall concentrate here on those aspects of it that are most relevant to atomic and subatomic physics.

Relativity theory is logically independent of quantum mechanics, and the two were developed separately. After the earliest version of quantum mechanics, which was not consistent with Einstein's relativity theory, was discovered, it took several more years to develop a form of quantum mechanics consistent with the requirements of relativity theory. The development led to many surprising results not easily anticipated from quantum mechanics without relativity, or from relativity without quantum mechanics. The results, such as the creation of matter, have been confirmed in experiments with subatomic particles, and have been essential in the understanding of the behavior of those particles.

Einstein, in what is called general relativity theory, showed it is possible to relate the observations made by any two observers, no matter how different the observers. However, for the purposes of atomic and subatomic physics, it is most useful to consider a restricted set of observers. This restricted set may be described as those related to one another in one or a combination of the following ways:

(a) being situated at different points in space;
(b) being at different times;
(c) moving at a constant velocity relative to one another;
(d) being rotated through some fixed angle relative to one another.

It is presumed that apart from these differences, the observers are otherwise identical, in that their instruments are made in the same way, and that they are able to carry out the same type of measurements. The reason for singling out this particular set of observers is that the laws of physics are found to be related in a simple way for all the observers of the set, whereas for other observers, the relations between the physical laws they will discover are more complicated. The set of relations between the observations of these different observers and between the laws they use to describe their observations is called special relativity theory, and was first clearly expressed by Einstein in 1905 for the physical laws known then. Einstein found that some of the laws had to be modified if they were to be simply related for different observers in relative motion. When these modifications were found to be true, physicists became convinced that all laws had to follow the relations Einstein had proposed, and this criterion is now used to restrict the laws that describe new phenomena. We shall see how this works in the case of quantum mechanics.

The principles of special relativity theory were first stated by Einstein in terms of measurements of position, and of the time of occurrence of events. However, these principles are somewhat more useful in the context of atomic and subatomic physics when stated in terms of energy and momentum, because those concepts are more generally applicable to atomic and subatomic systems. The general question that is answered by special relativity theory can be phrased in terms of the relationships between measurements of the energy and momentum of some physical system by the different observers described above.

Let us first consider the simplest kind of physical system we know, a single subatomic particle in free space. Assume that one observer has measured the momentum of this particle and obtained some definite value. According to Heisenberg's relation, this implies that the particle is equally likely to be found anywhere in space. It is then plausible that any two observers, differing only by their position in space, would obtain the same results for measurements of this system, and that is indeed what the relativity theory requires in this case. The result is different if the information about the single particle is not its momentum. For example, if one observer finds the particle to have a definite position, then another observer at a different point in space will find it to have another position, differing from that found by the first observer by the distance between the two observers. If the information about the particle is not one of these two extreme cases, the relation between the results obtained by the two observers displaced in space is more complicated, but can be expressed in terms of the mathematical quantity known as the Schrödinger wave function mentioned in Chap. III.

Consider next the measurements of a single stable particle by two observers at different points in time. This would seem to be more complicated, as the particle might change some of its properties between one time and the other. However, if the total energy of the particle has a definite value, as determined by one observer, and if this total energy does not change in time, then it turns out that all of the particle's properties will be the same for the two observers. In other words, the relativity principle in quantum theory requires that a particle with a definite energy should not change any of its observable properties with time. This is quite different from the situation in Newtonian physics, according to which an object can have a definite constant energy and nevertheless have complicated internal properties that change with time in a measurable way. But, the measurement of the energy of a quantum-mechanical particle forces all of its properties to become independent of time, as long as the energy does not change with time, just as the measurement of the momentum of the system forces all of its properties to be independent of space. For example, the spin direction of the particle must be constant in time. This is actually true not only for a single particle, but for any system in quantum mechanics which is in a definite energy state that is not degenerate.

At first sight, this would seem to involve a paradox, as it is hard to see how a smooth transition can occur from the situation in quantum theory to the Newtonian situation. The resolution of the paradox involves two factors. The first is that in quantum mechanics a system whose energy is not known precisely or whose energy state is one of several degenerate states, in general, will not appear the same at different times, but instead will have some properties that vary with time. The second is that for the large objects that approximately obey Newtonian mechanics, there are generally a very large number of possible energy values that are almost exactly the same. A measurement of the energy of the object is usually not precise enough to distinguish between these approximately equal energies, and therefore a large object will generally be in a state in which the energy is not precisely defined. In such a state, the value of various quantities other than the energy can vary in time which is in agreement with the observation.

The comparison of measurements carried out by observers that are relatively rotated in space through some angle is similar to the case of observers of different points in space. If the system being observed has a definite value of the magnitude of angular momentum for one such observer, then the different observers will see it as having the same magnitude of angular momentum, but with different values of the projection of the angular momentum along any direction. An interesting case is a system where the angular momentum is zero. Such a system will appear the same to all the rotated observers, i.e., it will be isotropic. This implies that such a system can have no directional properties associated with it, even if complicated internal motions occur in it. For example, if the system decays, it must do so isotropically, i.e., in such a way that any one of its decay products has equal probabilities of emerging in any direction in space. A system with zero angular momentum in quantum mechanics is therefore very different in its behavior from a similar system in Newtonian mechanics, which is, in general, not isotropic.

Systems that do not have definite total angular momentum, such as a particle with a definite linear momentum, appear differently to various rotated observers. The linear momentum appears to be rotated through the negative of the angle of rotation of the observers. However, it is generally the case that for an isolated system not acted on by external forces the energy of the system is the same for all rotated observers. It is possible to prove by mathematical analysis that this result is equivalent to the statement that the angular momentum of an isolated system is constant in time.

We consider next the most subtle application of relativity theory: to the case of observers which are in relative uniform motion. This is the case that Einstein treated in his 1905 paper, and the one which involves the greatest departure from Newtonian considerations. We consider again a single particle in empty space, which we imagine has been found to have a definite energy and momentum as measured by one observer. For another observer, moving with respect to the first, we should expect that the particle would be found to have different energy and different momentum, since each of these quantities depends on the velocity of the particle, and this velocity would be different as seen by the two observers. For example, the particle might be found at rest by one

observer, and so have zero kinetic energy and momentum, while
the observer in motion with respect to the first will see the particle
in motion, and so find nonzero values for both of these quantities.
Relativity theory provides the precise relation between the quantities measured by the two observers. At the same time, relativity
theory has led us to change our view of the relation between energy, momentum, and velocity as measured by a single observer.
The new relations are similar to those of Newtonian physics for
the case of objects whose velocity is much less than the speed of
light, but the relations become substantially different for objects
moving near or at the velocity of light, as is often the case for
subatomic particles.

The need for such a change is perhaps clearest for photons.
Since photons are the components of light rays, they travel at the
speed of light. Furthermore, careful measurements have shown
that the speed of light in a vacuum is always the same for all light
rays, and for whatever observer is measuring it. The direction in
which the light ray is moving does vary, depending on the motion
of the source of light and of the observer. If the relation between
energy, momentum, and speed of a photon were that of Newtonian mechanics, all photons would have the same energy and
momentum since they have the same speed. On the other hand,
the energy and momentum of photons vary in the way we have
described for light of different frequency. Therefore, the Newtonian relations cannot be correct for photons, and it is plausible
that they should also be different for other particles that are moving at almost the speed of light.

A clue to the correct relation for photons is the fact that the
ratio of the amount of momentum to the amount of energy is a
constant for all photons since both are proportional to the frequency of the light (according to the discussion in Chap. III). It
is therefore reasonable that this ratio involves the constant speed
of light, and, indeed, it is numerically just the reciprocal of the
speed of light. This suggests that for photons the velocity could be
defined in terms of the momentum and energy, rather than the
usual procedure of defining momentum and energy in terms of velocity and mass. Actually, this connection can be made more
rigorously by invoking a principle that Einstein took as the basis
of relativity theory. This is known as the relativity principle, which

states that the laws of physics should be the same for any two observers in uniform relative motion, so that it is impossible for any observer to determine whether he is at rest or in motion. It follows from the principle that the relation among velocity, energy, and momentum of any particle will be the same for all such observers. So if any one observer finds that the velocity of a photon is proportional to the ratio of momentum to energy, then all observers must find the same relation for this photon. But since a given photon is observed to have different energy and momentum by observers in relative motion, this implies that the energy–momentum–velocity relation must also be true for photons of various energy and momentum, and so is universally true.

The Doppler Effect

It is possible to use this result to infer how the energy and momentum of photons vary from observer to observer, a result originally obtained by Einstein through other methods. It is found that the energy of a photon as measured by one observer can be expressed in terms of both the energy and momentum as measured by another observer and the relative velocity of the two observers. Similarly, the momentum measured by one observer can be expressed in terms of those three quantities. For example, if the energy of a photon is measured by two observers, one of which is moving toward the source of the photon and the other not moving toward the source, then the photon energy, and also its frequency, will be greater for the moving observer. Conversely, an observer moving away from the source will measure the photon to have a lower energy and frequency than the one at rest with respect to the source (Fig. 14). By the principle of relativity, it must also be true that when the source of the photon is moving toward him, an observer will measure a higher frequency for the photon, and when it is moving away, he will measure a lower frequency than when the source and the observer are relatively at rest. In other words, the change in frequency depends only on the relative velocity of the source and the observer.

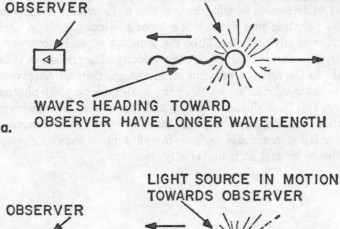

FIGURE 14. Doppler effect. Light waves emitted opposite to the direction of motion of a light source as in a. are stretched out, and so have larger wavelengths and lower frequencies when seen by an observer in that direction than they would if observer and light source were relatively at rest. When the waves are emitted in the direction of motion of the source, as in b., they are compressed, and so have shorter wavelengths and higher frequencies than for a source at rest.

These results are similar to some known for sound waves. If a fire truck is approaching you, the pitch, or frequency of the sound, of the siren is higher than of the same siren at rest. When the truck passes, the frequency suddenly drops, and when the truck moves away, the frequency is lower. This result is known as the Doppler effect, after the nineteenth century physicist Christian Doppler who first pointed out that this should occur for light as well. The effect is much more noticeable for sound than for light, because it depends on the ratio of the speed of the source to the speed of sound or to the speed of light, and the latter speed is 10^6 times greater. Nevertheless the light Doppler effect can be observed using a spectroscope and other precision instruments, and it agrees with the predictions of relativity theory. An important consequence of relativity theory, later discovered in the experiments with light, is the transverse Doppler effect which implies that the frequency of light will change even when the source is moving across the field of view, rather than toward the observer. In this case the frequency always decreases, and the effect is much smaller than the ordinary Doppler effect unless the source speed is very close to that of light. Because the amount of momentum of a photon is proportional to the amount of energy, similar results are found for the relation between the momenta of a photon as measured by relatively moving observers.

Once it is known how energy and momentum vary from observer to observer for photons, it is possible to extend these results to other particles that do not move at the constant speed of light. To do this, we note that according to the relativity principle the conservation laws of energy and momentum must be found valid by all observers in uniform relative motion if they are valid for any one such observer. Assuming this to be the case, it follows that the measurements of all forms of energy and momentum by different observers must be related in the same way as are those of a photon. We know this because the energy and momentum of photons can be converted into other forms of energy and momentum by processes such as Compton scattering. The relations between the quantities measured by different observers, either for photons or for other particles, are called Lorentz transformations, after a physicist of the nineteenth and twentieth centuries, who used them in a somewhat different context.

An important difference that occurs for particles whose speed can vary is that it is possible for them to be at rest as seen by some observer. In this case, the momentum of the particle would be zero. If the energy were also zero for a particle at rest, as suggested by Newtonian mechanics, then the Lorentz transformations would imply that energy and momentum would be zero for all observers, even those for whom the particle was in motion. Since this result is obviously incorrect, it must be that there is an extra form of energy, which does not vanish when a particle is at rest. This is not in disagreement with ordinary experience, provided that this extra amount of energy does not change in most circumstances. The energy that a particle has when it is at rest is called the rest energy. From the way rest energy is defined, it is an invariant quantity, the same for all observers of a given object.

The energy and the momentum of a moving particle can be calculated from the rest energy by the Lorentz transformations, and are proportional to the rest energy, providing that the rest energy is not zero, as it is for photons. The kinetic energy of the moving particle is the difference between this energy and the rest energy. When this kinetic energy is calculated for a particle that is moving slowly and compared with the expression from Newtonian mechanics it is found that the two expressions will approximately agree. The rest energy is related to the mass that appears in the formulas of Newtonian mechanics by the notorious formula $E=mc^2$, where E is the rest energy, m the mass, measured when the object is at rest, and c is the speed of light. This formula makes it possible to calculate the rest energy for an object whose mass is known, and this energy turns out to be very large in many cases. For instance, an object with a mass of 1 kilogram (kg) has a rest energy of about 3×10^{10} kilowatt hours (kWh), or about the amount of electrical energy used in New York City each year. For objects moving slowly compared to the speed of light, the rest energy is by far the largest part of the total energy. However, it is possible for the kinetic energy to exceed the rest energy if the particle's speed is greater than about 86% of light speed, quite common for subatomic particles.

Several points should be made about this rest energy. The first is that in any reaction among objects which remain the same, the masses of the objects do not change, and so neither do the rest energies. In such reactions, the large rest energies are therefore irrelevant because they are inaccessible, and the conservation of energy applies to the kinetic and potential energies alone, although the relation between kinetic energy and speed is somewhat different than in Newtonian physics. However, when the objects in the reaction can change character, as in many of the reactions among subatomic particles that we shall discuss, the rest energies will change, because different particles have different rest energies and the conservation of energy must include this rest energy.

A second point is that if we apply these ideas to a complex system such as an atom or a nucleus, the rest energy that must be related to the mass is not just the sum of the rest energies of the constituents but also includes any kinetic or potential energy that they may have while bound together. For example, the mass of a nucleus is equal to the total energy of the nucleus when it is at rest as a whole, divided by c^2. Since the nucleus as a whole can be at rest while its parts are in relative motion, this total energy will differ from the sum of the masses of the constituents by the total kinetic energy plus the total potential energy. If the nucleus is truly bound together, this extra energy will be negative, so that the mass of the nucleus will be less than the mass of all its constituents together. By measuring the mass of the nucleus, and the mass of its constituents, it is possible to determine how much kinetic energy should be generated when the constituents are brought together to make the nucleus. This method is sometimes used to predict which nuclear reactions can be used to produce usable amounts of energy.

The Relation Between Energy, Momentum, and Velocity

When we use the Lorentz transformation equations to express the energy and momentum in terms of the mass (or the rest energy) and the velocity, we find that the momentum is no longer proportional to the velocity, or the kinetic energy to the square of the velocity, except for slowly moving objects. Instead the dependence on velocity is such that the energy and momentum increase more rapidly with velocity than in Newtonian mechanics, and their values for any object whose rest energy is not zero become indefinitely large as its velocity approaches that of light. Since there is a limited amount of energy available to transfer to any object, this implies that no object whose mass is not zero can reach the speed of light, although they can come extremely close to that speed. Those objects that travel at the speed of light, such as photons, actually have zero mass, or zero rest energy. That is not unreasonable, because we have seen that such objects can never be brought to rest, as their speed is always equal to that of light. The theory of relativity therefore implies that two types of objects exist: those which always travel at less than light speed, and those which always travel at light speed. A third type of object which would always travel faster than light is also allowed by the theory, but is not yet known to exist. These new properties of energy and momentum for rapidly moving objects have been exhaustively tested by experiments involving subatomic particles and found to be accurate. There is no doubt that Newtonian mechanics must be replaced by relativistic mechanics in describing rapidly moving objects. It is truly remarkable that the relativity principle and the fact that the speed of light is a constant lead to correct conclusions about the relation between energy, momentum, and velocity for all objects in the world.

One other peculiar consequence of relativity theory should be mentioned, as it plays an important role in some behavior of subatomic particles. In general, the rate at which a process occurs for any subatomic particle decreases inversely as its energy increases. This means that if we compare the rates as measured by two observers, one of whom sees the particle at rest while the other sees it in motion, the rate will appear slower to the latter. Therefore, the half-life of an unstable particle in motion will be longer than that for the same particle at rest. In some cases of particles moving very nearly at light speed, this time slowdown effect can increase their half-lives millions of times.

Newtonian mechanics must also be modified because of quantum effects for subatomic particles. A correct description of these particles would therefore involve a version of quantum mechanics consistent with relativity theory. This was invented in the period from 1925–1950 by a number of physicists, and gives an accurate description of the properties of subatomic particles within certain limitations. The major limitation has been that the description does not determine precisely which particles—of the many types allowed by the theory—exist. However, relativistic quantum theory has predicted many unsuspected aspects of subatomic particles, such as the creation and annihilation of particles from each other, and has been used by physicists in their speculations on the nature of such particles. The details of relativistic quantum theory are still too mathematically abstract for simple exposition, but the results are not hard to understand as we shall see in the following chapters.

VI

The Atomic Nucleus, and How It Holds Together

Nuclear Sizes

The early measurements of alpha particle scattering by Ernest Rutherford and his co-workers, Johannes Geiger and Ernest Marsden, showed that the atomic nucleus is much smaller than the size of the electron orbits. Actually, nuclei vary in size depending on their mass. For all but the lightest nuclei, the volume is proportional to the mass, so that the density of matter in a nucleus has an approximately constant value, equal to about 10^{15} g/cm^3, or some 10^{15} times that of ordinary matter. This is another way of saying that the size of the nucleus is about 1 part in 100,000 of the size of the outer electron shells since we have seen that in ordinary solid matter, the outer shells of neighboring atoms are approximately in contact. If neutral matter, not containing electrons, exists, its density might approach the density of a nucleus, so that a quart of such matter would have a mass of 10^{12} tons. Such matter is believed to exist inside certain collapsed stars, called neutron stars. A star of this type with a mass equal to that of the Sun would be only about 10 km in diameter. A neutron star is in some ways like a single giant nucleus. But such neutral nuclei can only exist if they are large enough for gravity to play a role in holding them together. Ordinary nuclei cannot be electrically neutral for reasons we shall discuss, and cannot have a mass more than a few hundred times greater than a hydrogen atom without becoming unstable against fissioning. For this reason, the density of ordinary matter is determined by the size of the electron shells, and the nucleus remains hidden inside these shells, like a minute nut in a huge nutshell. The particles in a nucleus, which we shall soon discuss, are about as close together as they can be, similar to the atoms in a solid. Nuclei, unlike atoms, do not have large empty spaces in them.

Because of the need to describe nuclei by quantum mechanics, the shape of nuclei is not a simply defined concept. For all nuclei with zero-spin angular momentum, it is convenient to take the shape as spherical, since, as described in Chap. V, no measurements to reveal a deviation from this shape can be made. Nuclei with spins can have properties related to a nonspherical shape, and we shall mention these later. However, for many purposes, we can think of all nuclei as being spherical if we do not refer to measurements involving their spin.

The determination of the size of various atomic nuclei is an involved procedure and precise results have been obtained only since around 1950. Two methods have been used, both involving the electrical force acting between the positively charged nucleus and light, negatively charged particles. One method, developed by Robert Hofstadter and his collaborators, uses the scattering of high-energy electrons from various elements. The electrons are accelerated to energies of several hundred MeV by repeatedly applying electrical forces to them while they travel down a long tube, known as a *linear accelerator*. These high-energy electrons then are allowed to pass through targets made of various elements, and the fraction of the electrons that are scattered through various angles are measured. Electrons of such high energy are scattered hardly at all by the atomic electrons, but are scattered by the nucleus. If the nucleus were a mathematical point, the electrical force on an electron that approached very close to this point would become extremely large, and this force would generate the large change in momentum necessary to scatter the electron easily through large angles. On the other hand, if the nucleus has a nonzero radius, an electron approaching the center of nucleus within this radius will actually feel a weaker force than electrons remaining outside the nucleus. This means that relatively fewer electrons would feel large forces, and so be scattered through large angles, than would occur for a point nucleus. High-energy electrons are needed because of Heisenberg's relation. If an electron's position is well enough determined that it can be very near the center of a nucleus, it will have a correspondingly large uncertainty in momentum and in kinetic energy, which is possible only if the average kinetic energy is also high. From a precise measurement of the number of electrons scattered through various angles, the distribution of charge within the nucleus, which determines the electrical force exerted on an electron passing through the nucleus, can be inferred. To a good approximation, the charge distribution of most nuclei is found to be relatively uniform, and falls sharply to zero, so that it can be specified by giving just the radius of the nucleus.

The measurements of Hofstadter and others have shown that the radius of a nucleus is approximately 10^{-13} cm multiplied by the cube root of the atomic weight. Except for the few lightest elements this formula holds true over the whole Periodic Table. The *proton*, which is the nucleus of hydrogen, has a radius of about 7×10^{-14} cm, so even the simplest nucleus is not a point charge. The largest known nuclei, with atomic weights of about 250, have radii of 7×10^{-13} cm, and are about 10 times larger than the proton. This radius is still 70 times smaller than that of the innermost electrons of the atoms containing these nuclei, so the separation of electrons and nucleus remains definite, and finite nuclear size has little effect on the electron orbits.

There are, however, other negatively charged particles, much more massive than electrons, that can form atoms. One such particle is the muon, with a mass about 200 times greater than that of the electrons. The muon changes spontaneously into electrons and other particles in the relatively short time of 10^{-6} sec after it is produced. However, this is long enough for negative muons to be captured into orbits around nuclei. The lowest-energy orbit of a muon has a radius of only $\frac{1}{200}$ of that of the corresponding electron orbit around the same nucleus. This follows from the Heisenberg relation, because these orbits will have the same value of Δv, and the Δx, which determines the size of the orbit, varies inversely as Δp, which is $m\Delta v$. Therefore a more massive particle will have a smaller Δx, and hence a smaller orbit.

The muons' orbits are sufficiently small that these particles actually have a substantial probability of being found inside the nucleus. This is especially true in heavier atoms, where the nuclear radius is greater, and the orbits are smaller. Because of this, the force exerted by the nucleus of the muon, and therefore the energy of the muons' orbit, will be changed drastically from what it would be for a point nucleus. These energies can be measured accurately, by detecting the photons emitted when the muon shifts from one orbit to another, just as the energy of electron orbits is measured. From these measurements, it is possible to infer the size of the nucleus and the distribution of charge within the nucleus. These results agree accurately with those obtained from electron scattering, and leave no doubt that nuclei have finite sizes, varying between 10^{-12} and 10^{-13} cm.

The Constituents of Nuclei

Several important consequences follow from this observation, according to the principles of quantum mechanics. The first is that no electrons are contained in nuclei. The reason for this conclusion is simple. If an electron were confined in a region of 10^{-12} cm or less, then, by Heisenberg's relation, it would have a very large Δp, and an average kinetic energy of tens of MeV. Since the electron is confined, its potential energy would have to be more negative than this large kinetic energy is positive. But the electrical potential energy of an electron inside a positive nucleus of this size is nowhere near this great, being at most a few MeV. Furthermore, the electron scattering experiments, which would detect any potential energy of an electron in or near a nucleus, do not indicate any other strong potential. The conclusion that must be drawn is that electrons do not ordinarily exist in nuclei. Other arguments lead to the same conclusion. This result implies that when electrons are emitted from nuclei, as in the process of beta decay, the electrons are created just when they are emitted. This strange inference is the first indication we have seen of one of the central features of subatomic physics—the creation of particles from one another.

The absence of electrons from nuclei leads to another inference about what *is* in nuclei. It is known that for most elements, the nuclei may occur in different forms, with the same charge and different mass. These different forms are called *isotopes*. For instance, hydrogen nuclei occur in three forms, all with a charge of $+e$, but with masses of approximately one, two, and three times the atomic mass unit. This can be demonstrated by ionizing hydrogen atoms and deflecting the nuclei with a magnetic force. The force depends only on the charge, and is therefore the same for all three nuclei. However, the acceleration, and hence the deflection, depends inversely on the mass, and so will be different for the three isotopes. This method can be used to separate the isotopes of hydrogen or other elements from one another, and one such method was considered to separate uranium isotopes during the Manhattan Project. Isotopes of one element cannot be separated easily from each other by chemical methods because they contain the same number of electrons, in very similar orbits, and so behave quite alike chemically. But in some cases there are small differences in physical properties of atoms containing different isotopes, and these differences can be detected by living creatures which prefer one of the isotopes.

Since the isotopes of an element have the same nuclear charge, it is plausible to assume that they contain the same number of positively charged particles, and, because of the numerical value of this charge, that these particles are protons, or nuclei of the lightest hydrogen isotope. But to explain the variance in nuclear mass from isotope to isotope, it is clear that the nuclei must differ in their content of some other objects. In order to give the right mass values for different isotopes these objects should have no electric charge and approximately the same mass as the protons. Particles with such properties were discovered by James Chadwick in 1932, and called *neutrons*. Nuclei are made up of protons and neutrons, usually in similar numbers. For example, the nucleus of hydrogen with atomic weight two, known as the deuteron, has one neutron and one proton. Very heavy nuclei, such as uranium, tend to contain more neutrons than protons, for reasons that will be discussed later. The proton and neutron are called *nucleons*, because nuclei are made of them.

The Forces Between Nucleons

Another consequence of our observation is that nuclei are held together by nonelectrical forces. The fact that nuclei contain only positively charged protons and uncharged neutrons immediately shows that forces other than electrical ones are needed to hold them together, because no attractive electrical forces occur among such particles. For example, there is no electrical force between the proton and the neutron in a deuteron, yet the two remain within a few times 10^{-13} cm of each other indefinitely. In nuclei containing many protons, the electrical forces are repulsive, tending to break the nucleus apart rather than binding it together. From the size of nuclei, it is possible to estimate the strength of the force needed to bind them together, assuming that the measured size of nuclei corresponds to orbits of the protons and neutrons that are as small as allowed by Heisenberg's relation. It is found that the force binding protons and neutrons together in a nucleus must be from ten to one hundred times stronger than the electrical force between protons, at least over the size of a nucleus. This would also allow the electrical repulsion of the protons to be overcome by the stronger nonelectrical attraction. The greater strength of the nuclear force, compared with electrical forces, implies that the potential energy corresponding to this force is much greater than the electrical potential energy would be for two protons in a nucleus. Furthermore, the much greater mass of the proton and neutron compared to an electron implies that these particles have much less kinetic energy when confined to a nucleus than an electron would have. These two facts mean that the problem mentioned above, about confining electrons in a nucleus, does not occur for neutrons and protons.

The strength of the nuclear force indicates that it cannot be identified with any of the other forces known before 1930. The magnetic forces between protons tend to be even weaker than electrical forces, and gravitational forces between particles of the mass of neutrons and protons are some 10^{40} times weaker than electrical ones. There appears to be no conclusion possible other than that the nuclear force is an independent aspect of nature, with its own properties and characteristics.

One of these characteristics is that although the nuclear force is large when two nucleons are near one another as they are in a nucleus, it becomes very small when they are further apart, as for example are the two protons in a hydrogen molecule, which are about 10^{-8} cm apart. We know this because the properties of the hydrogen molecule are well described by the assumption that only electrical forces act among the protons and electrons contained in it. If there were a strong attraction between the protons at their distance in the molecule, we would expect the protons to come together. Similarly, if there were a repulsion much larger than the electrical repulsion, the molecule would tend to become much larger than it is. The conclusion we may draw is that the nuclear force is much smaller than the electrical force at a typical atomic distance. This conclusion is affirmed by experiments in which protons or neutrons are scattered from nuclei, in much the same way as we have described electron scattering from nuclei. In the proton scattering experiments, it is found that unless the proton approaches to within 10^{-12} cm or less of the nucleus the only detectable forces on it are electrical. However, at distances less than this, the nuclear force begins to act, and the protons are scattered much more strongly. When neutrons are scattered, only small magnetic forces act until the neutron gets within 3×10^{-13} cm, at which point the nuclear force becomes very strong.

These experiments indicate that the nuclear force varies much more rapidly as the distance between nucleons changes than does the electrical force, since the nuclear force is greater within the nucleus, and is much smaller at distances of 10^{-8} cm. For this reason, the nuclear force is described as short range, while electrical forces are called long range. The distance between two nucleons at which the nuclear force becomes large is known as the range of the nuclear force, and is approximately 3×10^{-13} cm. It should not be thought that the force is zero beyond this range; only that it is greatly reduced in value. The total force acting on a nucleon that is near or in the nucleus is the sum of the nuclear and electrical forces exerted on it by all the nucleons in the nucleus. For a proton, this total force is repulsive when the proton is beyond the range of the nuclear force, and tends to be attractive when the proton is within the range (Fig. 15). Because of this, it is difficult for a proton to approach a nucleus from outside; to do so, it must have enough velocity to pass through the region in which it is repelled, and come close enough to be attracted by the nuclear force. The kinetic energy required for a proton to be able to do this is about 1 MeV, if it approaches a light nucleus; more is required for a heavy nucleus. Actually, because of an effect peculiar to quantum mechanics called tunneling, the proton can reach the vicinity of the nucleus at a somewhat lower energy than this discussion would suggest. However, the result is qualitatively as

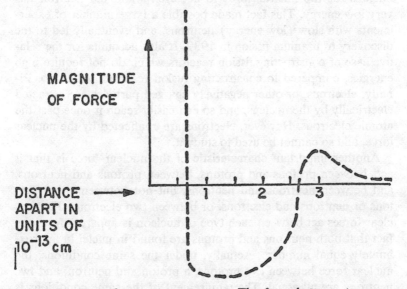

FIGURE 15. Force between two protons. The force between two protons varies in a complex way with the distance between the protons. When the distance is greater than about 2.5×10^{-13} cm, the force is almost entirely due to the electrical repulsion of the protons. As the distance decreases, the attraction due to the strong nuclear force becomes predominant. At still shorter distances below 5×10^{-14} cm, there is some indication that the nuclear force becomes repulsive. In the graph, a positive value indicates repulsion, a negative value attraction.

given here. The need for such relatively high proton energies to study the nuclear force led to the development of machines such as the cyclotron, which could accelerate protons to such energies. On the other hand, we shall see that this same requirement of a high energy for a proton, or any other positive charge, to be able to get within the range of the nuclear force, has the desirable consequence that it allows elements with light nuclei, such as carbon and oxygen, to continue in existence, even though they could in principle undergo transformations into other nuclei, when they collide with one another.

The situation is different for a neutron, because there is no electric repulsion, and therefore a neutron can come close enough to a nucleus for the nuclear force to act even when the neutron has very low energy. This fact made possible a large number of experiments with slow (low-energy) neutrons, and eventually led to the discovery of uranium fission in 1939. It also accounts for the relative ease of constructing fission reactors which do not require high energies, compared to constructing fusion reactors, which do. Finally, electrons, or other negatively charged particles are attracted electrically by the nucleus, and so can easily reach it once past the atomic electrons. However, electrons are unaffected by the nuclear force, and so cannot be used to study it.

Another important characteristic of the nuclear force is that it acts between protons and protons, between protons and neutrons, and between neutrons and neutrons, but not between either protons or neutrons and electrons, or between two electrons. That nuclear forces act between each type of nucleon is apparent from the fact that both neutrons and protons are found in nuclei in approximately equal numbers. Actually, under the same conditions, the nuclear force between two protons, a proton and neutron, and two neutrons are all equal. The requirement of the same conditions is essential, and more subtle than might appear. Two protons are two of the same type of particle, and satisfy the exclusion principle with one another just as two electrons do. Similarly, so do two neutrons. However, a proton and neutron are different particles, so no exclusion principle is observed between them. Because of this, two protons with the same spin projection cannot be found at the same point in space, while a neutron and proton with the same spin projection can. The equality of neutron–proton forces with proton–proton and neutron–neutron forces is meaningful only in situations that can occur for all three systems. One illustration of this is the ground state of the deuteron which is a neutron–proton system with no orbital angular momentum, and with nucleons which have equal spin projection. For two protons, no such state can occur by the exclusion principle, and indeed there is no bound nucleus of two protons at all, not because the nuclear forces are different from the neutron–proton force, but solely because the exclusion principle does not allow the one bound state to be filled.

The equality of the neutron–proton, proton–proton, and neutron–neutron nuclear forces is in sharp contrast to the electrical forces, which act only among charged particles, be they nucleons or others. This equality is referred to as the charge independence of nuclear forces, and indicates that there is some relation between the neutron and proton, in spite of their different electric charge. This relationship was the first clue to a remarkable set of similar relationships that occur between many of the subatomic particles discovered in recent years. The elucidation of these so-called internal symmetries has been a major element in the progress of subatomic physics since 1900, and will be discussed in detail in Chap. VIII.

The study of the physics of nuclei has gone in two distinct directions since the discovery of nuclear force and the neutron clarified the general qualitative properties of nuclei. One aspect of nuclear physics has investigated the nuclear force itself in order to learn as much as possible about its properties and origin. This study has for the most part examined the force between two isolated nucleons whose behavior is much simpler than that of many-nucleon systems. It has led to a detailed understanding of the nuclear force, but, more significantly, it has led to the realization that this force is only one aspect of a much richer array of phenomena generally known as *strong-interaction* physics. These phenomena involve not only nucleons, but also other subatomic particles, and include other changes in particle properties than the change in momentum that we describe as a force.

The other aspect of nuclear physics has involved the study of the behavior of nuclei themselves, concentrating on both their structure and their transformations into one another. In this study, a few properties of the nuclear force have been used to analyze the properties of nuclei in terms of their constituent neutrons and protons. However, for reasons we shall soon see, this approach has had limited success. Instead, most information about nuclei has come either from experiment or from simple nuclear models that are relatively independent of the details of the force between nuclei. In this respect nuclear physics is somewhat similar to the physics of complex atoms and molecules, where it is also difficult to calculate precisely the properties of the atoms from a knowledge of the constituents and the forces acting between them, simply because so many constituents are involved.

Some Properties of Nuclei

A basic property of any nucleus is its mass. It is possible to measure the mass of a nucleus by the technique used to separate isotopes, called mass spectrometry. An ion of an atom containing the nucleus is passed through a region where a uniform magnetic force acts on it. The ion travels through part of a circular orbit while in this region. The radius of the circle depends on the ratio of charge to the mass of the ion. Since the mass is almost entirely that of the nucleus, ions with different nuclei will traverse different orbits. By measuring the radius of the orbit, the ionic mass can be determined. A small correction for the mass of the electrons contained in the ion can then be made to ascertain the mass of the nucleus alone.

When this procedure is applied to an element as it is found in natural materials, there will in general be several different isotopes, each of which will traverse different orbits. The masses inferred for the different isotopes through the measuring process are therefore unequal, and usually differ from one another by approximately, although not exactly, the mass of a neutron. This is further evidence for the hypothesis that isotopes differ in the number of neutrons they contain. The relative amounts of different isotopes contained in natural materials varies greatly from element to element. For example, 99.98% of natural hydrogen is the isotope of mass 1, and 0.02% the isotope with mass 2. On the other hand, chlorine is a mixture of two isotopes with mass approximately 35 and 37 that of a proton in a proportion of 3 to 1. The atomic weight of an element as determined by chemical methods is an average of the weights of the different isotopes multiplied by the proportion of that isotope found in the natural element. Because of this, the atomic weight may not be an integer multiple of that of hydrogen, and indeed is not for chlorine and many other elements. But the atomic weight of an individual isotope is much closer to an integral multiple of that of hydrogen, although here also there are deviations.

It is also possible to find relations among the masses of different nuclei by applying the conservation of energy to transformations in which one nucleus changes into another and then using the Einstein mass–energy relation. A combination of these methods has been used to determine with great precision the masses of many isotopes of all of the elements.

We shall always refer to the mass and energy of the ground state of the nucleus. As with an atom, the nucleus has excited states in which one or more of the nucleons has a higher energy than in the ground state, and in which the nucleus as a whole therefore has a higher energy and mass. These excited states are unstable, and, like their atomic counterparts, make transitions to lower-energy states of either the same, or other, nuclei by emitting photons or other particles. By measuring the energies involved in these transitions, the energy differences between the ground state and many excited states have been determined for most nuclei, and are typically about 1 MeV. This is a small fraction of the rest energy of the nucleus, which is approximately equal to the atomic weight multiplied by the rest energy of the proton (about 940 MeV). Nuclei are often referred to by the name of the element and the nucleon number, which is the sum of the number of neutrons and protons. Thus uranium-235 means the isotope of uranium containing 235 nucleons.

Another important property of nuclei, the charge, can be determined in several ways. One of these methods involves the determination of the energy of the innermost electrons by measuring the energy of the X rays emitted when an outer electron drops into a hole in this shell vacated by an electron that has been removed.

The inner electron energy is a simple function of the nuclear charge, and so a measurement of the energy will determine the charge. Alternatively, the nuclear charge determines the scattering of high energy, charged particles through small angles by an atom containing the nucleus, provided that the energy of the charged particles is high enough to penetrate the electron clouds, but low enough that the nuclear force is unimportant. Thus a measurement of this scattering will also fix the charge of the nucleus. By the use of these and other methods, it is straightforward to determine the charges of various nuclei. It is found that the charge of different isotopes of the same chemical element are the same, which is to be expected, since the nuclear charge of the neutral atom is the same as the number of electrons, and the chemical properties of the element are determined by the number of electrons.

A knowledge of the nuclear charge determines the number of protons in the nucleus, since there are no negatively charged particles present to cancel any of the protons' charge. The number of neutrons present is a bit harder to determine. Whereas the charge of the nucleus, or of any system, is the algebraic sum of the charges of its constituents, this is not the case for most other nuclear properties. In particular, the mass of a nucleus is not just the sum of the masses of all the protons and all the neutrons making up the nucleus. This is even true in the deuteron, whose mass is less by about 0.1% than the sum of the masses of a proton and a neutron. Heavier nuclei typically have masses that are about 1% less than the sum of the masses of the constituent protons and neutrons. The reason for this was mentioned earlier. We know that any system that stays together must have a binding energy. That is, its kinetic and potential energy together must be negative. But then the total energy of the system will be less than the sum of the rest energy of its constituents, since the total energy includes the kinetic and potential energy as well as the rest energy. If it were possible to synthesize a nucleus by slowly bringing together all the nucleons that compose it, the difference between the sum of the rest energies of these nucleons and the total energy of the nucleus would be given off in some form, either as photons or other subatomic particles that would be created in the process. The mass of system is related to its total energy by the Einstein formula. Since the total energy of the nucleus is less than

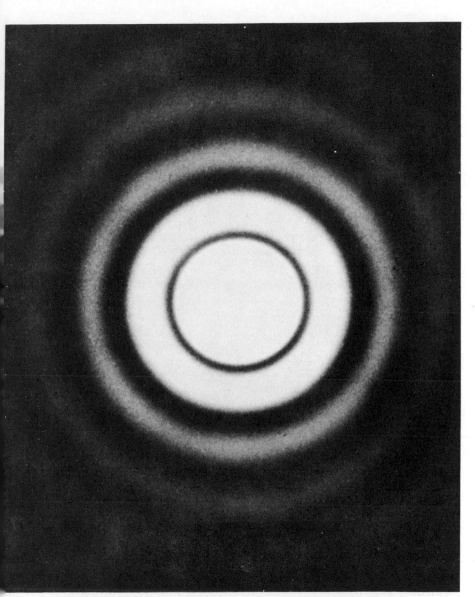

PLATE 1. Diffraction of visible light by a small hole in a screen.

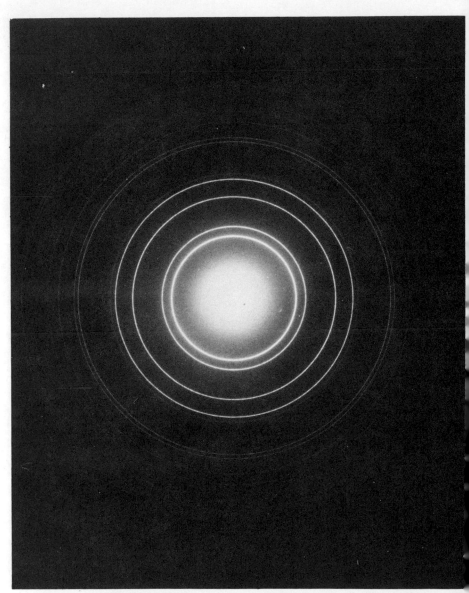

PLATE 2. Diffraction of electrons by an aluminum crystal. The similarities in this pattern with those in Plate 1 indicate that a wave is associated with each phenomenon.

PLATE 3. Fermi National Accelerator Laboratory. An air view of the synchrotron at Fermilab in Batavia, Ill. The tall building in the foreground is sixteen stories high. The accelerator ring is below ground, but its outline may be seen. The ring is four miles in circumference.

PLATE 4. Creation of electron—positron pairs by a photon. In this bubble-chamber picture, an invisible photon enters from the right, and transforms into a positron, which gives the large spiral track on the lower half and an electron, which is the long curving track on the upper half. The curvature is due to a magnetic force acting on the particles. The other lines in the picture are unrelated to this process.

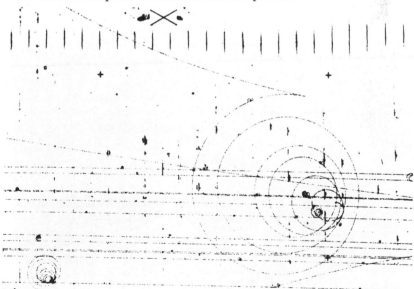

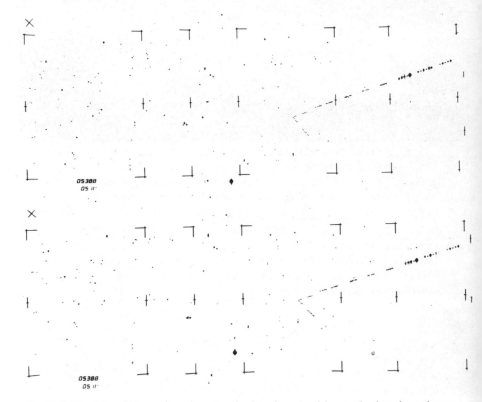

PLATE 5. Neutrino interactions in a spark chamber. In this spark-chamber picture, a mu-neutrino from the left-hand side hits a neutron, producing a muon and proton. The picture may be compared with Plate 8, showing a similar event in a bubble chamber.

the sum of the rest energies of its constituents, the mass of the nucleus will be less than the sum of the masses of the neutron and protons that make it up. Another way to say this is that the binding energy of the nucleus, being negative, is equivalent to a negative mass, which reduces the mass of the nucleus below the sum of the masses of its constituents. We can go even further and recognize that the kinetic energy of the nucleons in the nucleus—being positive—makes a positive contribution to the total mass, while the potential energy coming from the nuclear force and other forces acting between the nucleons makes a negative contribution, because the over-all force is attractive and this corresponds to a negative potential energy.

In view of this, we can translate the statement that the mass of an average nucleus is 1% less than the sum of the masses of its constituents into a statement that the binding energy of a nucleus is about 1% of the sum of the rest energies of its constituents. This may be compared to the situation in a hydrogen atom, where the binding energy of the electron is only 0.0025% of its rest energy. This is so because the particles in a nucleus are much closer together than the electron is to the nucleus of an atom and because the nuclear force is stronger than the electric force.

For a uranium atom containing over 200 nucleons, the mass will be less than the sum of the masses of its constituents by about two nucleon masses. Therefore, if we tried to estimate the number of neutrons by just using the mass of the nucleus, we would underestimate the number by two. Instead, we estimate in the following way. For light nuclei, such as oxygen, the mass is close to an integer times the neutron or proton mass. Therefore, since the number of neutrons must be an integer, we can determine this number just from the mass of the nucleus. For all heavier atoms, the mass of the oxygen nucleus is always within 0.1% of some simple fraction of the mass of the heavier nucleus. By dividing this fraction into 16 (the number of nucleons in oxygen), we can determine the number of nucleons in the heavier nucleus. The number of neutrons can then be found by subtracting the proton number given by the nuclear charge. The complexity of this procedure is an example of a problem that becomes much more serious when the possible constituents of subatomic particles are considered where the binding energies may be comparable to these rest energies, and where the number of constituents cannot be inferred from the masses.

The results of this procedure give a definite value for the proton and neutron number of each nucleus. These values are now known for several hundred nuclei with proton numbers ranging from 0 to 105 and neutron numbers ranging from 0 to about 150. We compare nuclei with a fixed number of nucleons, but divided differently between neutrons and protons. Several general features of these nuclei are apparent.

(1) In light nuclei (those with less than about 40 nucleons), the lowest rest energy comes for nuclei with approximately equal numbers of protons and neutrons.

(2) For nuclei with more than 40 nucleons, the lowest energy occurs in nuclei with more neutrons than protons, and the neutron–proton ratio increases as the total number of nucleons increases. A minimum value of the energy for a given number of nucleons is important, because there is a tendency for nuclei with various numbers of protons and neutrons, but with the same number of nucleons, to convert into the nucleus with the lowest energy by various processes such as *beta decay*. Because of this tendency, it is usually the case that of all nuclei with the same number of nucleons, only the nucleus with lowest energy is stable, the others being radioactive. Therefore, the stable nuclei that compose ordinary matter are those with approximately equal numbers of protons and neutrons for light nuclei, and those with somewhat more neutrons than protons for heavier nuclei.

(3) The total binding energy divided by the number of nucleons is a measure of how strongly the average nucleon is bound in the nucleus. This average binding energy increases rapidly from zero (for the lightest isotope of hydrogen) to about 7 MeV/nucleon in the helium isotope containing four nucleons. It then increases slowly, with minor fluctuations, to value of about 8.7 MeV/nucleon in the nuclei containing 56 nucleons such as iron-56. After that, it decreases slowly to about 7.5 MeV/nucleon in the heaviest known nuclei (Fig. 16). The average binding energy for a particular nucleus is an important determinant of what nuclear transformations the nucleus can undergo.

It is possible to understand these features of nuclei from very simple properties of the nuclear force. Since this force is the same

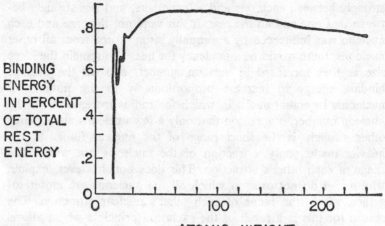

FIGURE 16. Average binding energy of nuclei. The average binding energy of a nucleus is small for very light nuclei, and reaches a maximum of about 0.9% of the total rest energy for nuclei containing around sixty nucleons. Such nuclei are the most stable, and the average binding energy decreases slowly beyond that number of nucleons. The curve has been smoothed out and does not show the fluctuations in binding energy between isotopes and other nuclei containing approximately equal numbers of nucleons.

between all nucleons, we might expect that the binding energy of all isotopes would be the same. However, this does not take into account that, because of the exclusion principle, different states become available if one of the protons in a nucleus is replaced by a neutron. We have seen this already in the deuteron, which has a state available to it that is impossible for a two-proton or a two-neutron system. Similarly, a nucleus with equal numbers of protons and neutrons usually has states available that could not exist if one of the protons became a neutron, or vice versa. This implies that the nucleus with equal numbers of protons and neutrons tends to have the lowest energy because it has the option of being in one of the extra states. This accounts for the stability of the nuclei with equal proton and neutron numbers in light nuclei.

The fact that the average binding energy is roughly the same for all but the lightest nuclei indicates that the nuclear force does not act equally between all particles in a nucleus, but instead acts

strongly between each one and a few others, and less strongly between that one and all the rest. If this were not the case and each
nucleon was influenced by an equally strong force from all other
nucleons, there would be a tendency for nuclei to remain the same
size as they increased in nucleon number, and for the average
binding energy to increase proportionately to the number of
nucleons. In actual nuclei in which the radius increases with the
nucleon number, one reason that only a few nucleons attract each
other strongly is the short range of the nuclear force. In the
heavier nuclei, only a fraction of the nucleons are within this
range of each other's attraction. This does not, however, explain
why nuclei do not exist in which all the nucleons are closer together, within the range of each other's nuclear attraction. The
reason for this is a result of the exclusion principle which allows
at most four of the nucleons in any nucleus to be in the same orbital state, i.e., two neutrons and two protons. Any other nucleon
will be in a different orbital state than these four. If the nuclear
force has the property of being strong only for nucleons in the
same orbital state, this would account for the facts mentioned.
This turns out to be the case, because of the way the nuclear force
originates. This behavior of the nuclear force is known as saturation. Empirically, we can see this from the fact that helium-4,
which contains just the four nucleons that can occupy the same
orbit, already has an average binding energy of 7 MeV/nucleon,
whereas neighboring nuclei, with more or less particles, have
appreciably lower average binding energy. Of course there must
be some attraction between nucleons in different orbits or else all
nuclei would break up into units of four, but this extra attraction
is relatively small. This then explains why nuclei increase in size
as the number of nucleons increase. The exclusion principle forces
the extra nucleons to occupy orbits which tend to be larger, and
this is not compensated by an increase in the force exerted on each
nucleon.

Finally, the departure from equality in the number of protons
and neutrons in heavy nuclei is to be understood as a result of the
electrical repulsion of the protons. Unlike the nuclear force, this
electrical repulsion is not short range, nor does it saturate. As a
result, even though the electrical repulsion between two protons is
small compared to the nuclear force attraction, when a nucleus

contains many protons, all repel each other electrically by similar amounts which builds to an over-all repulsion of each proton that can be comparable in size to the nuclear attraction by the few nucleons in the same orbital state. As a result, the binding energy of a nucleus with more neutrons, which do not have any electrical repulsion, can be greater than that of a nucleus with equal numbers of neutrons and protons, even though the neutrons must occupy orbital states whose binding energy from the nuclear force is less. However, the over-all binding is less than would occur if there were no electrical repulsion operating. As a result, the average binding energy in heavy nuclei is decreased below its value for the nuclei around iron. This leads to the possibility of such heavy nuclei breaking up into smaller nuclei, in the process called fission.

The interplay of the lesser nuclear binding of the states that extra neutrons must occupy and the lesser electrical repulsion in nuclei with an excess of neutrons determines the criterion of stability. For nuclei with many more nucleons than any yet known, this criterion would result in stability with a small fraction of the total being protons. However, such nuclei would have relatively low average binding energy compared to the known lighter nuclei, and therefore be susceptible to spontaneous breakup into such nuclei.

Nuclear Transformations

One of the important things we have learned about nuclei is that they can be transformed into one another under suitable circumstances. This can either happen spontaneously, as it does in various sorts of radioactive decay, or as a result of exposing the nuclei to outside stimuli, such as energetic beams of nucleons, photons or other particles. In either case, the transformations of most interest in nuclear physics can be thought of as rearrangements of the protons and neutrons into different configurations, or of a change of neutrons into protons, but in which no extra particles of the type that undergo strong interactions with the nucleons are produced. Later, we will consider reactions that produce extra particles of that type; such reactions are best studied in the framework of individual particle transformations, rather than nuclear transformations.

The spontaneous breakup of a nucleus into two or more parts was first observed among the natural radioactive isotopes, in the form of *alpha decay:* the emission of a helium-4 nucleus, or alpha particle, by a heavy atom which leads to another chemically distinct atom. We know that the residual atom will have two less protons and two less neutrons in the nucleus, since these constitute the helium-4 nucleus. The atom will also lose two electrons because of the reduced nuclear charge, but this occurs after the alpha decay. Two quantities generally characterize alpha decay: the energy of the emitted alpha particle and the half-life for the decay. It is interesting that whereas the energies of the alpha particle do not vary much, say between 4 and 10 MeV, the half-lives vary by many orders of magnitude, from 10^{-7} sec to 10^{+10} years. These facts have been explained by a theory of alpha decay, originated by George Gamow and others in 1928, which I will discuss below.

More recently, the spontaneous breakup into two approximately equal parts has been observed for a number of nuclei with masses greater than 230. This process is known as spontaneous fission, as it is quite similar to the fission process induced by neutrons involved in nuclear energy generation. Actually, both spontaneous fission and alpha decay can occur in the same type of nucleus, and according to Gamow's theory their relative probability depends sensitively on the nuclear charge and binding energy.

The most fundamental requirement that must be satisfied in order that a specific nuclear breakup can occur is that energy be conserved. If the initial nucleus with mass m is at rest, its total energy is just the rest energy mc^2. Suppose the nucleus breaks up into several nuclei with various masses. The total energy of these final nuclei cannot be less than the sum of their rest energies; it could be greater if the nuclei emerge from the breakup with some kinetic energy. Note that any binding energy of the particles within one of the final nuclei will be counted as part of the mass of that nucleus. There can be no binding energy or potential energy between the various final nuclei because when they emerge they must travel far enough from each other to be detected separately, say at least 1 cm, and at this distance all forces between nuclei are negligible. Therefore, a necessary condition for the occurrence of a spontaneous nuclear breakup is that the mass of the nucleus be greater than the sum of the masses of all the final nuclei. One might imagine that this condition could be avoided if the initial nucleus were in motion when it breaks up, since it will then have some kinetic energy as well as its rest energy. However, because relativity theory tells us that if a transformation cannot happen for a single particle at rest it also cannot when that particle is in motion, this motion is of no consequence, and the condition inferred for the nucleus at rest is general. The same condition, that the rest mass of the initial object is greater than the sum of the rest masses of the decay products, holds for all decays of subatomic particles.

By consulting a table of masses, or of binding energies of nuclei, it is then possible to infer when a nuclear breakup can occur. We have seen that the average binding energy increases until a nucleon number around 60 is reached, and then decreases slowly beyond that. This implies that those nuclei with less than 60 nucleons generally will not break up spontaneously, as their mass will be less than the sum of the masses of any set of lighter nuclei containing the same total number of nucleons. On the other hand, nuclei with 200 nucleons or so have less average binding energy than those with 60 nucleons, and so can break up into several such nuclei. Furthermore, the mass of a nucleus with more than 150 nucleons is usually greater than the mass of the nucleus containing two less protons and two less neutrons plus the mass of an alpha particle. Therefore, the heavy nucleus can break up by alpha decay.

The energy released in an alpha decay or spontaneous fission is just the difference between the binding energy of the initial nucleus and the sum of the binding energies of the final nuclei. This energy emerges in the form of the kinetic energy of the nuclei produced in the transformation. In alpha decay, the kinetic energy goes almost completely to the alpha particle, because it and the heavy residual nucleus must carry away equal linear momentum and the lighter alpha particle will then have most of the kinetic energy. Because the average binding energies of heavy nuclei decrease only slowly as the number of nucleons increases, the kinetic energy released in an alpha decay or a spontaneous fission does not vary much from nucleus to nucleus, in agreement with the observation mentioned above. Furthermore, the kinetic energy will be a small fraction of the binding energy of the initial nucleus. In other words, most of the binding energy of the initial nucleus remains as binding energy of the product nuclei. In alpha decay, the kinetic energy released is only 0.5% of the total binding energy, which itself is only 1% of the rest energy. In spontaneous fission, the kinetic energy release can be 10% of the total binding energy, but still only 0.1% of the rest energy. Although some other nuclear reactions may release as much as 0.6% of the rest energy of the nucleons involved, most of the energy in a nucleus cannot be obtained by nuclear reactions. Nevertheless, these reactions release roughly one million times more energy per particle that participates than do typical reactions involving the atomic electrons, mainly because nuclear binding energies are so much larger than electron energies in atoms.

The extreme variability of the lifetimes of nuclei for alpha decay can be understood in terms of the relative ease or difficulty with which the alpha particle escapes from the nucleus. We can imagine that the alpha particle is already present in the nucleus before the decay, because the kinetic energy of a bound alpha particle is not large. If the decay is to be energetically possible the separated alpha particle and residual nucleus must have a lower rest energy than the initial nucleus. However, in order to escape, the alpha particle must pass through a region of space just outside the nucleus in which the nuclear force is very small and the electrical force is larger (see Fig. 15). In this region, the electrical repulsion produces a large positive potential energy, and the energy of the alpha particle is much higher than its energy either inside or far outside the nucleus. This region then acts as a barrier to the emergence of the particle from the nucleus. It is analogous to a marble inside a small hole on top of a large hill. The marble can obtain a lot of kinetic energy by rolling down the hill. But in order to do this, it must first obtain enough energy to get out of the hole. In Newtonian physics, this can only happen if the marble is given energy from an outside source. In quantum mechanics, a particle in a similar situation can emerge spontaneously by a process known as tunneling (Fig. 17).

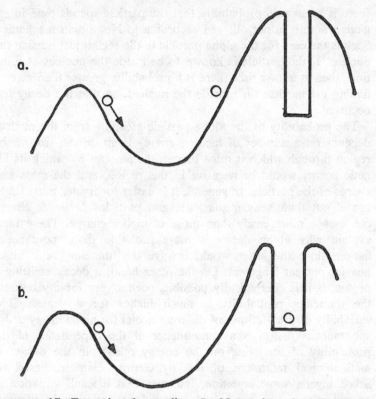

FIGURE 17. Example of tunneling. In Newtonian physics, a ball rolling from the top of a low hill does not have enough energy to reach the top of a higher hill, as in a., and so cannot enter the hole in the hill, even though its energy at the bottom of the hole would be equal to its energy outside. In quantum physics, as indicated in b., the ball can enter the hole, by making a direct transition from outside to inside. This process is fancifully known as tunneling.

The essence of tunneling is the possibility of a particle moving from one point where it has some potential energy to another point, where its potential energy is lower, through a region of limited extent, in which its potential energy would be so high that if energy were conserved, its kinetic energy would have to become negative. While this is impossible in Newtonian mechanics, in quantum mechanics the position and the kinetic energy cannot be determined together because of Heisenberg's relation. As a result, there is a nonzero probability that the particle spends time in regions where it is not allowed according to Newtonian mechanics. Such is the case for the alpha particle in the region just outside the nucleus. If the particle is known to be inside the nucleus at some time, then at a later time there is a probability greater than zero of finding the particle far outside the nucleus, so that the decay has occurred.

The probability of the alpha particle escaping from the nucleus depends on a number of factors, among them, in the size of the region through which it must tunnel, the amount by which its kinetic energy would be negative in this region, and the mass and charge of the particle. In general, it is easier for lighter particles to tunnel out than heavier particles, and particles of lower charge can escape more easily than those of higher charge. These facts explain why alpha decay is more probable than spontaneous fission, since the latter would involve the tunneling of a much heavier nuclear fragment. On the other hand, a decay emitting a proton, if it is energetically possible, occurs very rapidly because the tunneling probability is much higher for a proton. The variability of the lifetimes of different nuclei for alpha decay or for spontaneous fission is a consequence of the dependence of the probability of tunneling on the energy release in the decay. A mathematical treatment of this by George Gamow, based on Schrödinger's wave equation, indicates that a small variation in energy release can produce a tremendous variation in lifetime. This predicted relation is in agreement with observations of many nuclei.

All of the known nuclei containing more than 210 nucleons can decay via alpha emission or spontaneous fission. It is energetically possible for alpha decay to occur for some nuclei with less than 208 nucleons, but the lifetimes involved would be extremely long, even compared to the 10^9 years or so of some existing alpha decays. Therefore, practically speaking, these lighter nuclei may be considered as stable against alpha decay. The atomic number of lead is among the highest of such alpha-stable nuclei, and so lead is the eventual end product nucleus of the chain of many radioactive decays of heavier nuclei. As the number of nucleons increases beyond 250, the lifetimes for spontaneous fission and for alpha decay become quite short. Such nuclei will therefore have decayed long before the present time, even if they were synthesized in earlier stages of the universe. It is possible to produce such nuclei in the laboratory, so that their properties can be studied, but they do not persist long enough to accumulate in any significant quantities. Recently there have been speculations that certain nuclei containing about 300 nucleons could have substantially greater average binding energies than some of the known lighter nuclei. This would make these nuclei relatively more stable against alpha decay and spontaneous fission than the lighter nuclei, and these superheavy nuclei could have fairly long half-lives. No such nuclei have yet been discovered, but efforts have been begun to produce them, by the methods discussed below.

An observation of some importance is that the alpha particles emitted by a particular nucleus have only a few distinct energies, rather like the photons emitted in atomic transitions. This indicates that nuclei, like atoms, have quantized energy levels, and that the different energies of the alpha particles correspond to having either the initial nucleus or the final nucleus in various of their energy levels. As in atomic transitions, it is possible to get information about the energies, angular momentum, and other properties of the nuclear levels by studying the energies of the emitted alpha particles. Another consequence of the discrete energies observed in alpha decays is that no other particles are emitted together with the alpha particles in the decay. This is so because if another particle were emitted, it would share the total available energy with the alpha particle in an arbitrary ratio, and alpha particles would be observed with a continuous range of energies from zero to the maximum allowed by energy conservation. Such phenomena are seen in fission, where extra neutrons are often emitted, and in beta decay, where a particle known as a neutrino is emitted along with the electron.

Beta Decay

In beta decay, a nucleus changes into another nucleus containing the same number of nucleons, but with a charge greater by

+e. In addition, a beta particle, or electron, is found to emerge, as we have seen in Chap. II. Since the electron has a charge of $-e$, the charge will balance. On the other hand, a measurement of the angular momentum of the initial and final nuclei involved in beta decay indicates that the two always differ by an integer multiple of $\hbar$. Since the electron has a spin of ½ $\hbar$, this would imply that angular momentum is not conserved, unless another particle is also emitted with the electron and residual nucleus. Further evidence for this comes from careful observations, which show that the residual nucleus moves in a direction that is not precisely opposite to the electron. Since the initial nucleus is essentially at rest, and so has no linear momentum, the conservation of linear momentum would require these directions to be opposite, again unless a third particle is involved. Finally, as mentioned above, the emission of an extra particle which can share the available energy with the electrons allows the electron energy in beta decay to vary continuously. This extra particle must be uncharged, since the charge balances between the electron and nuclei. It must also have a spin of ½ $\hbar$, in order to conserve angular momentum. The existence of such a particle in beta decay, called a neutrino by Enrico Fermi, was hypothesized in 1930 by Pauli, and proved finally in the 1950s, by the detection of neutrinos in absorption experiments. Neutrinos, like photons, have zero rest mass.

The simplest example of nuclear beta decay is that of a free neutron into a proton, electron, and a neutrino. This occurs with a half-life of ten minutes, so that neutrons are relatively unstable objects. In order for this decay to be possible, the neutron mass must be greater than the proton mass, which it is by about 1.5 MeV. Many of the beta decays of complex nuclei can be thought of as the beta decay of one of the neutrons into a proton which remains trapped in the residual nucleus. In stable nuclei, the nuclear forces act to lower the energy of the neutrons below what an extra proton would have, and the beta decay becomes energetically impossible. If the neutron energy in the nucleus is lowered enough below that of a proton, it becomes energetically possible for a proton to decay into a neutron. This involves the emission of a neutrino and a *positron,* or positive electron, a particle we have not yet introduced, but will discuss presently. Some nuclei are known in which this kind of proton beta decay occurs.

Nuclei that undergo beta decay are very common throughout the Periodic Table of elements, indeed, there are some known with almost every mass number. The half-lives for beta decay are never as short as for some alpha decaying nuclei, and a typical value would be several minutes. Some information about the properties of nuclear states can be obtained from beta decay, but the analysis is too complicated to describe. Perhaps of more interest is the insight that a study of beta decay has given to the properties of subatomic particles. Since neither electrons nor neutrinos are present in nuclei, they must be created in the process of beta decay. A detailed theory of how this occurs will be discussed in Chap. X, in the general context of subatomic particle transformations.

Nuclear Reactions

While some information about nuclear transformations has been obtained by experiments with naturally radioactive nuclei, much more understanding has come from a study of transformations induced by exposing the nuclei to external probes consisting of various particles that collide with the nuclei. We have already discussed the information about nuclear radii obtained from the scattering of electrons by nuclei. In this section we shall concentrate on the phenomena observed when neutrons and protons collide with nuclei. It should be kept in mind that the situations being discussed are somewhat artificial, in that such collisions do not normally occur to most of the nuclei present in the matter on Earth. However, such collisions between nuclei are quite common in most stars, where they are responsible for the way energy is produced.

In collisions of neutrons with nuclei, since the neutron is uncharged, it has no problem in reaching the neighborhood of the nucleus, even if its energy is very low. When the neutron is close enough to the nucleus that the nuclear force can act upon it, one can think of the collision as having produced, at least for a short time, a new nucleus, containing one extra neutron, called the compound nucleus. The probability of this happening can be expressed as a cross section for the absorption of neutrons by the original nucleus, just as we have done for the absorption of light by the atom. A typical cross section for neutron absorption is 10^{-24} cm², a number called 1 barn, to indicate that it is relatively large. In some cases, however, especially for neutrons of very low energy, the cross sections can be many thousands of times larger than this.

The compound nucleus that is formed is generally in an excited state, rather than the ground state. This is because the added neutron, instead of having about 6 MeV of negative binding energy, has a small or large positive kinetic energy. Therefore the state of the compound nucleus that is formed is something like 6 MeV higher than the ground state. This state is usually highly unstable, and will rapidly break up in one of several ways, including the re-emission of a neutron, of a charged particle such as a proton, or an alpha particle, or by emission of a photon. These processes tend to occur in light elements. When they occur, yet another nucleus is formed, which may be different from the nucleus originally hit. Sometimes the new nucleus is still unstable, and a series of nuclear transformations occur until a stable nucleus is finally reached. If the cross section for the original neutron capture is high, and if an intense source of neutrons is available, say from a fission reactor, it is possible to use this process to transmutate elements in significant quantities. Except in the special case of plutonium production, discussed below, this has not been done in quantities large enough to replace natural sources of the stable isotopes of the element. However, it has been possible to produce by neutron capture radioactive isotopes of various elements that are useful for various industrial, scientific, and medical purposes.

In nuclei with 230 or more nucleons, a new process can occur after neutron absorption. The compound nucleus fissions into two nuclei, each about half the original size, with a total kinetic energy of 200 MeV, representing the difference in binding energies. The process is called induced fission. One can think of this as the spontaneous fission of the excited compound nucleus which generally has a much higher probability than the spontaneous fission of the ground state of the same nucleus, as the excited state has about 6 MeV higher energy. This extra energy is enough in some cases to allow the fission fragments to get through their mutual electrical barriers described above for alpha decay. While all nuclei can be made to fission by neutrons of sufficiently high energy, a few isotopes such as uranium-236 will fission when produced by low-energy neutrons hitting the parent nuclei uranium-235. These parent nuclei are radioactive, but their half-lives are so long that enough of them are present on Earth to be used as targets for neutrons.

After the fission occurs, the two residual nuclei are still radioactive, as they contain many more neutrons than protons, whereas a stable nucleus in that mass range would contain more equal numbers. The residual nuclei therefore decay rapidly, sometimes by emitting neutrons which are observed as part of the fission process. It is the occurrence of neutrons with the fission process that makes a chain reaction possible. Under certain conditions, the neutrons produced when one nucleus fissions can induce fission in another nucleus. The continuation of the process leads to the fissioning of a very large number of nuclei in a comparatively short time which is known as a chain reaction. If this did not occur, a new neutron would have to be furnished to initiate each fission, at an exorbitant cost in energy. This induced fission is the basis of some nuclear explosives, as well as of the generation of energy in fission reactors. The neutrons produced in the fissioning of a large number of nuclei become a source of neutrons that can be used for further studies of neutron-induced reactions. Note that because the free neutron is unstable and transforms in about ten minutes into a proton, electron, and neutrino, it is not possible to accumulate neutrons to be used as we choose.

Fission is not the only process that can occur after neutron absorption in heavy nuclei. For some compound nuclei, such as uranium-239, which is formed by low-energy neutron absorption in uranium-238, the probability of fission is still small, and this compound nucleus transforms by beta decay into an element with charge 93, called neptunium. This in turn transforms by beta decay into element 94, plutonium, still with 239 nucleons, as the nucleon number remains fixed in beta decay. Plutonium-239 is important because, like uranium-235, it can be induced to fission by low-energy neutrons, and a chain reaction can take place. On the other hand, a chain reaction cannot occur in uranium-238, because of the small induced fission cross section. By exposing uranium-238 to neutrons produced in the induced fission of uranium-235, it is possible to transform most of the uranium-238 nuclei into plutonium by the above process. This has the effect of increasing the amount of fissionable material available, because each fissioning uranium-235 nucleus produces an extra neutron which can produce a plutonium nucleus from a uranium-238 nucleus. A practical application of this process is the breeder reactor, now being considered as a source of electrical energy. Plutonium is now produced in amounts of many tons per year in ordinary reactors, and is stored for eventual use. The value of the energy that can be extracted from an ounce of plutonium is approximately $150, or about the value of an ounce of gold. Therefore, the large scale transmutation of uranium-238 into plutonium is a realization of the alchemist's dream, at least in monetary equivalent.

The sequential absorption of neutrons has been used to produce nuclei with higher and higher numbers of nucleons. Nuclei with as many as 260 nucleons have been made in this way, in quantities up to 1 g. These nuclei eventually beta decay and produce nuclei with more protons than the original target nucleus. Nuclei with proton numbers of up to 100 have been produced in this way. Beyond this, the nuclei have no isotopes that live long enough to be used as targets for neutrons. However, still heavier elements have been produced by other methods, discussed below.

The whole set of elements containing more than 92 protons are called transuranic elements. Because their lifetimes are all short compared to the age of the Earth, they are not found on Earth naturally, and awaited synthesis for their discovery. It has been speculated that all the nuclei with more than about 60 nucleons were produced by absorption of up to 180 neutrons by iron nuclei in a very short period of time during stellar explosions known as supernovae. Some evidence for this comes from the abundances of the various isotopes in the Universe, which agree roughly with what this model suggests. Direct evidence that such large scale transformations of nuclei can take place has been found in nuclear explosions, where up to 20 neutrons have been absorbed by uranium nuclei in a short time, producing heavy transuranic elements.

Reactions between positively charged particles and nuclei differ somewhat from those of neutrons. If the positive particle has an energy below a few MeV, the electrical repulsion between it and the nucleus will prevent it from getting close enough to the nucleus to be influenced by the nuclear force. This statement is only approximately correct because of the possibility of tunneling which can proceed into the nucleus as well as out from it, even when the kinetic energy is too low to surmount the barrier. However, the probability of tunneling is small until the kinetic energy is at least a few per cent of the value needed to overcome the barrier. Because of this, a low-energy proton or other positive charge projected toward the nucleus scatters from it mainly by the electrical force, just as an electron would.

A charged particle whose energy is high enough to overcome the electric repulsion of the nucleus will reach the vicinity of the nucleus, and will form a compound nucleus, much as a neutron does. The same sorts of phenomena then occur as in neutron-induced reactions. However, the charged particle reactions are of separate interest for several reasons. One is that unlike neutrons, charged particles can be moved about by electric and magnetic forces. This makes it possible to measure and to change their energies in more precise ways than can be done with neutrons. Therefore more accurate measurements of some of the quantities in charged particle reactions are possible than for neutron reactions.

A second reason for interest in these reactions is the possibility of collisions between two heavy nuclei. Such reactions are made possible by removing the electrons from a heavy atom and accelerating the resultant positively charged nucleus to a high energy by electric forces. If such a heavy nucleus is given enough energy to overcome the electric repulsion of another nucleus, a collision between the two will form a compound nucleus containing as many protons and neutrons as the sum of the colliding nuclei. This compound nucleus may contain many more nucleons than any naturally occurring nucleus. For example, some of the heavier transuranic elements have been produced by collisions between lighter transuranic elements and boron nuclei that have been stripped of their electrons. There are also proposals to produce some of the hypothetical semistable superheavy elements by collisions between very heavy stripped nuclei such as copper, and uranium nuclei. This could produce compound nuclei with over 300 nucleons, which could then decay to the metastable ground states.

Finally, charged particle reactions are of interest in connection with the process known as nuclear fusion, the source of energy which allows most stars to shine and which will perhaps someday be used as a source of energy for human activities.

We have noted that for nuclei with less than 60 nucleons the average binding energy increases as more nucleons are added. For example, the deuteron has a binding energy of about 2 MeV, while the alpha particle has a binding energy of 28 MeV. Therefore, if two deuterons can be combined to form an alpha particle, about 24 MeV of energy is released. This is almost 1% of the total rest energy of the particles involved. It is much less energy per atom than in uranium fission, but represents substantially more on a per nucleon, or per unit weight basis. The process involving the union of two lighter nuclei to produce a heavier nucleus, or sometimes one heavy and one still lighter nucleus, is called fusion.

The main problem involved in accomplishing fusion is that of giving the charged particles enough energy for them to overcome their mutual electrical repulsion. This can be done readily enough for a small number of charges by accelerating them by electric and magnetic forces, but it is not a useful way of obtaining energy in large quantities. A reasonable machine can accelerate 10^{16} protons/sec to a few MeV. A small fraction of these protons would induce fusion reactions if the protons were absorbed in a target made of deuterium. This reaction would produce about 10 MeV per fusion, but this would generate a total power release much less than the 10 kWh that would be needed just to run the machine.

A method that has instead been considered, and which actually occurs in stars, is to heat the substance containing the nuclei to be fused to a very high temperature. The temperature of a body is related to the average kinetic energy of the particles that compose it. In order for the average kinetic energy to be as high as 1 MeV, a body would have to be raised to a temperature of 10^{10} deg, or 500 times the temperature at the center of the Sun. However, this is unnecessary, because even at lower temperatures some fraction of the particles will have a kinetic energy far above the average, and the fusion reaction can occur to some extent for particles with less energy than that needed to surmount the electric barrier through tunneling. Consequently, it is thought that at a temperature of 10^8 deg or so, enough nuclei could undergo fusion—through a reaction such as two deuterons forming a helium-3 nucleus and a neutron—that energy would be released in substantial amounts. The main problem involved in accomplishing this is that of keeping the particles near each other long enough for this to happen, since the high kinetic energies and temperatures correspond to high pressures that tend to make the particles fly apart. Efforts to avoid this by confining the particles with magnetic forces have not yet proven successful, but work along these lines continues. In stars, the particles are kept together by their mutual gravitational attraction, but this is not feasible on Earth because not enough particles are available to bind one other gravitationally at the high temperatures required.

Nuclear Structure

The information obtained from nuclear transformations as well as from the properties of individual nuclei have been combined to give a reasonably accurate description of the internal structure of nuclei. This description of course must be in terms of quantum mechanics, since we have seen that the sizes of nuclei are such that the restrictions coming from Heisenberg's relation are relevant. Furthermore, nuclei have additional complications compared to atoms. One is that in nuclei there is no central body that acts as the source of most of the force on each nucleon. Instead, we have seen that the force on a particular nucleon in the nucleus is mainly exerted by a few nearby nucleons. As a result, the motion of each nucleon is rather more complicated than the motion of an electron in an atom. Another complication is that the force acting between two nucleons depends on more aspects of the nucleons than just their distance apart; for example, on their relative velocity. Furthermore, neutrons and protons, like electrons, carry an intrinsic angular momentum or spin, of $\frac{1}{2}\hbar$, and the nuclear force also depends on the relative orientation of these spins.

This also serves to complicate the calculation of the energies and other properties of nuclear states. It might, therefore, appear that there might be little in common between the nuclear states and the atomic states we have described in Chap. III.

In spite of this, similarities do exist between the states of nuclei and those of atoms which are summarized in a description of nuclear states known as the independent-particle model. In this model, the nucleus is taken to be a collection of nucleons moving under the influence of an over-all force, rather than under forces exerted between pairs of nucleons. The rationale for this is that the individual nucleons move rapidly within the nucleus, and over a short period of time of some 10^{-22} sec each nucleon is successively near to and then far from each other nucleon. If we consider the average behavior over a period of time that is longer than this 10^{-22} sec, the force exerted on each nucleon will be a kind of composite average of the forces exerted on it by all of the other nucleons. This average will be essentially the same for each nucleon, since they all interact successively with one another. This procedure of replacing the results of many individual interactions by an average effect is a common one in physics, and is often used to deal with situations in which it would be too difficult to calculate the behavior according to individual interactions. A similar procedure is sometimes used in atomic physics to describe the forces of the electrons upon one another, which gives a correction to the force on each electron due to the nucleus.

The precise form of the average force exerted on each nucleon is not important for determining the properties of the nucleus, and since this force is just an approximate description anyway, knowing its precise form would not be interesting. Only certain general features of the average force are relevant to the properties of the nucleus. One is that the potential energy corresponding to the force is taken to be symmetric under rotations, just as is the electric potential energy in atomic physics. This has the consequence that degenerate states exist among the nuclear energy levels, and leads to a shell-type structure of the nuclear energy levels similar to, but not identical with, that occurring in atomic physics.

Another important aspect is that the average potential energy of a nucleon depends not only on the location of the nucleon within the nucleus, but also on the spin direction of the nucleon and the orbital angular momentum of the nucleon. In particular, a nucleon with its spin and orbital angular momentum pointing in the same direction has a lower energy than one with spin and orbital angular momentum pointing in opposite directions. Both the existence of this so-called spin-orbit interaction and the rule about its magnitude are somewhat empirical notions, introduced originally to make the independent-particle model agree with observation. However, measurements of the interaction between individual nucleons done in scattering experiments indicate that a spin-orbit-type potential energy occurs in that case, and it is therefore plausible that the average potential energy of a nucleon obtained by adding together a number of two-nucleon potential energies also should have such a dependence. Actually, a similar type of spin-orbit potential is known to occur between electrons and the nucleus as a consequence of the theory of relativity. In that case, however, the size of this potential is much smaller than the usual electrostatic potential, except for the inner electrons in heavy atoms, where the two can be comparable in size.

As a result of the spin-orbit potential acting on nucleons, the energy of a nuclear level depends not on its orbital angular momentum or its spin angular momentum, but on the total angular momentum, obtained by adding the two. In other words, a complete set of quantities to describe a nuclear state would be the energy, the magnitude of this total angular momentum, the projection of the total angular momentum, and the magnitude of the orbital angular momentum. The description of the energy levels must be expressed in terms of these quantities rather than the ones we used in atomic physics. We note that since the orbital angular momentum is always an integer multiple of $\hbar$, and the spin of a nucleon is a half-integer multiple of $\hbar$, the total angular momentum of a nucleon will always be a half-integer multiple and can never vanish. On the other hand, for a nucleus containing an even number of nucleons, the sum of the angular momenta of all the nucleons can be zero, and usually is zero in a nuclear ground state.

Because of the saturation property of the nuclear forces, the average potential energy of a nucleon is approximately the same no matter how many nucleons are in the nucleus. This makes it possible to determine a set of energy levels for each nucleon in some nucleus that will be similar in energy, and in the various angular momenta, to those for each nucleon in another nucleus with a different number of nucleons. Since nucleons have a spin of $\frac{1}{2}$, they satisfy the exclusion principle, and thus two protons cannot occupy the same state, nor can two neutrons. As a result, as extra nucleons are added to a nucleus, they must occupy higher-energy states, which generally have higher angular momentum as well, just as do the electrons in atoms. This result helps to explain some of the observed properties of nuclear levels. For example, helium-4 is a very tightly bound nucleus whereas helium-5 or lithium-5, which would be obtained from helium-4 by adding an extra neutron or proton, do not exist as bound nuclei at all. The reason for this is that in helium-4 the two protons occupy the lowest-energy level in which each nucleon has total angular momentum of $\frac{1}{2} \hbar$. Therefore, the two protons in helium-4 have the two possible projections of the total angular momentum. Similarly, the two neutrons have the two spin projections of the lowest-neutron-energy level. There is no bar to having a neutron and proton in the same level, because they are different particles. However, if an extra neutron or proton is added, it cannot go into the same level as the other two because of the exclusion principle, and so must occupy a higher level. In this higher level there would be very little force acting on it, because, as we have seen earlier, most of the nuclear force comes from nucleons in the same orbit. Therefore, the extra nucleon is not bound to the others, and the nucleus will not hold together. This might suggest that no nuclei heavier than helium-4 should exist, in contradiction to nature. However, if two protons or two neutrons are added to helium-4, the resultant nucleus *is* bound, because the two extra nucleons exert an attractive force on each other as they can ocupy the same state.

By combining the exclusion principle and the rules for degeneracy of levels with a specific total angular momentum, it is possible to determine which nuclei will have nucleons in the same level and which will require some nucleons to be in higher levels. Those values of the neutron number or proton number at which the next nucleon would have to enter a higher-energy level are analogous to closed shells in atomic physics. The corresponding nuclei are significantly more tightly bound than the succeeding nuclei. The neutron and proton numbers for which this happens are called magic numbers, and were first recognized empirically from energy measurements. Later, the values of the magic numbers were calculated by the method indicated. The magic numbers are 2, 8, 20, 28, 50, 82, 126, 184, etc., for both neutrons and protons. Nuclei with magic numbers (and closed shells) of both neutrons and protons are especially tightly bound relative to their neighbors, for example, helium-4, oxygen-16, and lead-208. The superheavy nucleus with 126 protons and 184 neutrons would also be doubly magic, and that is why it is believed that this nucleus might be relatively stable.

This independent-particle model of nuclei also predicts the angular momentum and certain other properties of the nuclei at or near the magic numbers of protons and neutrons, and these predictions have generally been found to be true. It also makes predictions about some of the excited states of nuclei which should have properties related to the ground states of nuclei containing more nucleons as these extra nucleons would have to fill the higher orbits. These excited states of nuclei have been observed by scattering various particles from the nuclei, which can sometimes change the target nucleus to an excited state with a corresponding energy loss by the projectile. By measuring how much energy the projectile particle loses, the energy difference between the ground and excited nuclear states is measured, just as in the Franck–Hertz experiment in atoms. The excited nuclear states are unstable, just as are excited atomic states, and decay to the ground state, of either the same or another nucleus by emitting photons, alpha particles, or other particles. The precise decay mode depends on the detailed properties of the nucleus, and, occasionally, several alternate decay modes occur for the same excited state with various probabilities. In any case, the decay of excited nuclear states follow the same probability laws as atomic decays, or alpha decay, and are characterized by a half-life. The half-lives for decays involving photon emission are shorter than those for other decays, since no barrier penetration is involved— the photon is unaffected by either the nuclear force or the electrical force.

While the independent-particle model has been fairly successful at describing many of the properties of the ground states and excited states of nuclei, there are other aspects of nuclei with which it fails. One is the depiction of the formation of compound nuclei that occurs in many nuclear reactions. When a compound nucleus is formed in a collision, the energy of the projectile particle is divided up among the nucleons in the target nucleus, rather than being transferred to a single nucleon. Since this requires strong interactions among all the nucleons in the nucleus, it is not surprising that description in terms of independently moving nucleons would not explain it accurately. Other aspects of nuclei not well represented by the independent-particle model include the fact that some nuclei appear to deviate substantially from a spherical shape in favor of a cigar shape. This deviation has been recorded through measurements of the nuclear charge distribution, and the independent-particle model does not, at least in its simple form, predict accurately the magnitude of these shape deviations.

Other models of nuclei have been put forward to account for the features not well described by the independent-particle model. While some of these models are fairly successful, it remains the case that the properties of nuclei are less well understood than the properties of the electrons in atoms. This is not because any fundamental principles are missing from our description of nuclei, but rather because the simplifying factors present in atoms, such as the relative weakness of the forces and domination by the force exerted by the nucleus on the electrons, are not as valid within the nucleus. Nuclei are examples of the type of systems with which physicists have the most trouble—those with many interacting components, none of which can be safely neglected.

In spite of this, there exist at least qualitatively correct explanations for all of the observed properties of nuclei, and these explanations can be related to, if not always derived from, what we know about the individual nucleons that constitute the nucleus. In response to this situation, and in the belief that the individual nucleons would form a more fundamental and more tractable subject for study, the interest of many physicists has gradually shifted from nuclear physics to what has come to be called elementary particle physics. We shall see that this expectation has not been completely satisfied by the realities of elementary particle physics.

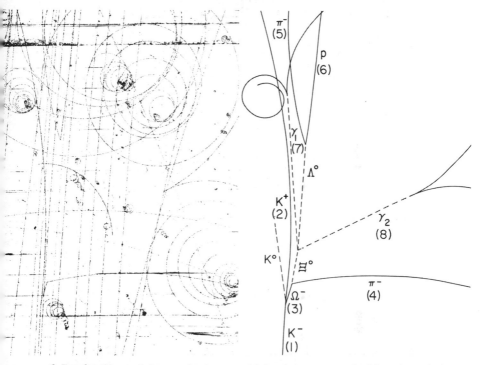

PLATE 6. Production and decay of an Ω^- particle. A K^- meson incident from below hits a proton producing an Ω^- particle, together with a K^+ meson and an invisible K° meson. The Ω^- decays into Ξ° and a π^-. The Ξ° then decays into a Λ° and π°. The π° decays into two gamma rays, which are detected when they convert into electron—positron pairs. The Λ° decays into a proton and π^-. The presence of the various neutral particles is inferred from measurements of energy and momentum of the visible charged particles. This picture was the first evidence for the Ω^- particle.

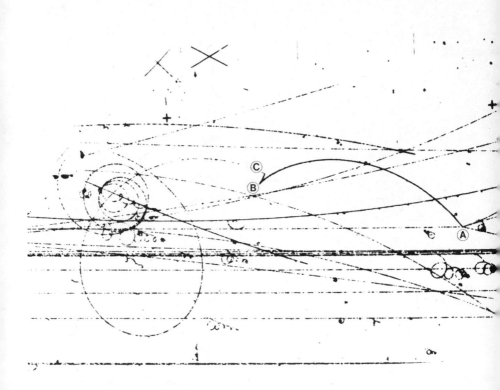

PLATE 7. Sequential decay of a pion, into a muon, into a positron. In this bubble-chamber picture, a positive pion is produced at A. It travels to B where it decays into a positive muon. The short muon track goes to C, where the muon decays into a positron, which is seen in the spiral track. Neutrinos are also produced at B and C, but remain invisible.

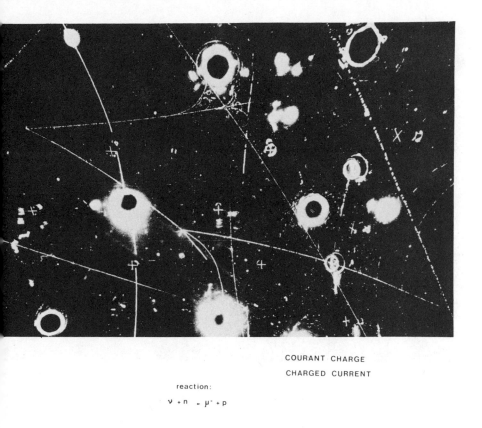

COURANT CHARGE
CHARGED CURRENT

reaction:

$$\nu + n \rightarrow \mu^- + p$$

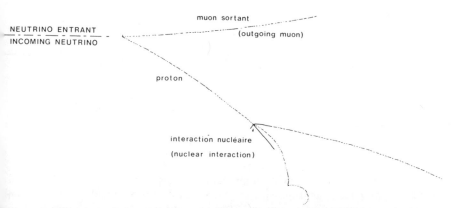

NEUTRINO ENTRANT
INCOMING NEUTRINO

muon sortant
(outgoing muon)

proton

interaction nucléaire
(nuclear interaction)

PLATE 8. Production of a muon by a neutrino. A mu-neutrino incident from the left-hand side hits a neutron in a nucleus. The neutrino converts into a negative muon while the neutron changes to a proton which later hits another nucleus, disintegrating it. This is an example of a charged current interaction of a neutrino.

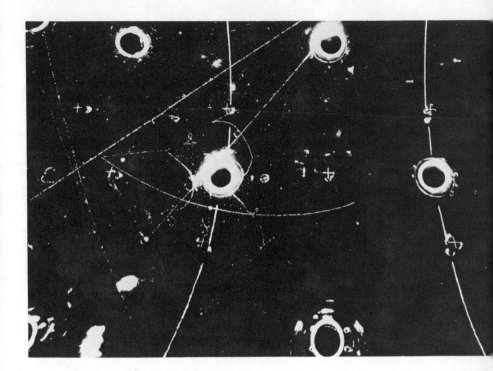

COURANT NEUTRE

NEUTRAL CURRENT

$$\nu_\mu + n \longrightarrow \nu_\mu + \pi^- + $$

gamma de pion neutre

(gamma from neutral pion)

unseen ν_μ

negative pion

NEUTRINO ENTRANT

INCOMING NEUTRINO

interaction nucléaire $\quad \pi^- p \to \pi^0 n$

(nuclear interaction)

proton (stop)

PLATE 9. Neutral current neutrino interaction. In this bubble-chamber picture, an incident neutrino hits a neutron, which transforms into a proton and a negative pion, while the neutrino remains a neutrino. The negative pion later hits a nucleus and converts to a neutral pion, which decays. The proton eventually stops in the bubble chamber because it has lost its kinetic energy. It is the nuclear interaction and the stopping that allow the hadrons to be distinguished from muons, which would not do either.

VII

Experimental Techniques in Subatomic Physics

In order to study subatomic particles, a number of novel techniques have been developed to give particles higher kinetic energies than are readily available from natural sources, and to detect the collision products of such high-energy particles with matter. In this context, the phrase "high energy" refers to kinetic energies of 100 MeV or more. Such energies are needed, according to Heisenberg's relation, in order to probe the inner dimensions of subatomic systems. These energies are also necessary to allow the production of most of the subatomic particles not found in ordinary matter. Particles with kinetic energies of 100 MeV or more are not common in the Universe. They do sometimes occur in what is called cosmic radiation, a stream of high-energy protons that continually hits the Earth from space. Some information about high-energy particles has been obtained through the study of cosmic radiation. However, the intensity of this radiation on Earth is too low for very detailed analysis, and it has been necessary to invent devices, such as synchrotrons and linear accelerators, to accelerate protons and electrons to high energy, in order to study their reactions.

The other major innovation in experimental subatomic physics has been the development of devices that detect high-energy particles, and measure and record some of their properties, such as momentum and electric charge. The need for new devices to do this is a consequence of the very small values of the quantities characterizing a single subatomic particle, especially the short times that these particles are available for measurements. These times are short both because some particles are *unstable,* that is, decay quickly into other particles, and because the high speeds of the particles remove them quickly from the measuring instruments. In spite of these difficulties, it has been possible to construct detectors that can reliably react to particles which spend as little as 10^{-11} sec in the detector. In many cases these detectors present a vivid picture of a series of subatomic processes that we believe is a faithful image of what actually happens.

The development of these experimental techniques has been an integral part of the progress of twentieth century physics. Without them, it would have been difficult or impossible to create or verify the theories that we use to describe subatomic systems. While the inventors of experimental methods are often unsung heroes outside their profession—and sometimes even inside it—the relevance of all theoretical ideas to natural phenomena rests ultimately on the application of the inventions of these instrument makers. It is especially important to remember this because the greater ease of describing developments in theoretical physics can easily be distorted into the conclusion that such theoretical developments are the essential part of physics. Time and again the introduction of new experimental techniques has led to results not previously anticipated by any theory, and it is the interplay of such experimental and theoretical advances, in a kind of intellectual leapfrog that has provided much of the excitement of twentieth century physics.

Particle Accelerators

The subatomic particles that emerge from atomic collisions in discharge tubes, or the other apparatus in which they were first found, have kinetic energies no greater than a few thousand eV.

Even those particles produced in nuclear transformations such as beta decay or fission generally have energies less than several million eV per particle. As a result of this limitation, around 1930 physicists turned to the construction of devices that would give higher energies to subatomic particles. In the forty-five years since then, the energies at which these devices operate have increased by a factor of a million, a factor comparable to the factor between the energy change in ordinary atomic transitions and that in nuclear transitions.

These devices, known under the general name of particle accelerators, all operate on charged, stable particles, including protons and electrons. This is so because the only practical way that has been found to accelerate particles is by the application of electric and magnetic forces whose action is most pronounced on charged particles. Furthermore, it is difficult to make and maintain the unstable particles in sufficient quantities to accelerate them to high energies. The general principles on which particle accelerators operate are fairly simple. A constant electric force applied to a charged particle will accelerate the particle, i.e., increase its energy, in proportion to the distance that the charge moves while the electric force acts on it. Therefore, to increase the energy as much as possible, the particle should move through a long distance, or equivalently, be acted on for a long time. However, it is impractical to have large electric forces that extend over great distances because such forces tend to affect all the material in them. Two methods have been used to deal with this problem. One is to confine the motion of the particle to a limited region and allow the electric force to act over and over on it, always in the direction in which the particle moves. This can be done by the use of a magnetic force in addition to the electric force. Such a magnetic force cannot increase the energy of the particle by itself, but it can make the particle move in a closed path, so that an electric field that extends along this path can act on the particle. This is the basic principle used in all circular accelerators, beginning with the cyclotron and leading to the modern synchrotrons.

In accelerators of this type, the charged particle travels in almost circular orbits under the influence of the magnetic force, with the orbital radius slowly increasing as the particle's energy is increased by the electric force. As long as the particle moves slowly compared to the speed of light, it takes the same time for it to make each revolution. Therefore, it can be arranged to have an electric force act on the particle at equally spaced intervals, in a direction such that the speed of the particle always increases, by having the electric force change its direction in the same period that the particle revolves. This method worked well in the early cyclotrons which were used to accelerate protons and alpha particles to energies of tens of MeV. It would not work well for electrons, whose velocity approaches that of light when the kinetic energy is 1 MeV or so. When this occurs, the period of revolution begins to vary considerably from turn to turn, and the particles are sometimes accelerated and sometimes decelerated by the electric forces. The same thing happens to protons with energy of many hundred MeV. This occurs because at high speeds the mass of the particle increases from its rest mass, and the acceleration produced by a given magnetic force decreases. One way to compensate for this effect is to vary the period of the electric force in such a way that it remains equal to the period of revolution of the particles, even when the latter is changing. This is done in machines known as synchrocyclotrons, which can therefore accelerate protons up to many hundreds of MeV. Incidentally, the orbits in a cyclotron, or other particle accelerators, are so large that they can be very well described by Newtonian physics, and the restrictions due to Heisenberg's relation are unimportant.

Still another adjustment is necessary to accelerate protons to higher energy in a practical way, and this is done in a machine called the *synchrotron*. The increase of the radius of the particle's orbits in a cyclotron would eventually require huge magnets inside which the particles would move. Instead, it is convenient to confine the motion to an orbit that is approximately fixed in size. This can be done by increasing the magnetic force in step with the increased mass of the particle, so that the radius of the orbit remains fixed. Thus in a synchrotron, the electric and magnetic forces both vary with time in such a way that the particle always traverses approximately the same orbit. When the speed is low, the time taken to traverse the orbit decreases continuously as the energy increases. However, when the speed approaches that of light, the particle cannot move any faster, and the period of revolution also becomes constant so that the electric force need not change with time by very much. This approach can be used for both electrons and protons because the effects of relativity are compensated. Synchrotrons have been built for electrons with energies up to about 10 GeV (billion electron volts) and for protons of energies up to 400 GeV. The latest of these, at the Fermi National Accelerator Laboratory (Fermilab) in Batavia, Ill., has a diameter of more than a mile, but the protons are confined to a relatively small tube, only a few inches in diameter (Plate 3).

The discussion given above is very idealized in that different protons in the same accelerator do not all have the same energy, and careful accelerator design is necessary to ensure that acceleration will occur for almost all particles regardless of their initial energy or direction. The accomplishment of this has made the study of such particles possible, because in addition to the need for high energy, it is important that a sufficiently large number of particles attain this energy so that experiments can be carried out with them. Existing circular accelerators are able to give a high energy to 10^{12} or more particles each second. While this is a large number of particles, it corresponds to a very small total mass. In fact, one can estimate that less than one gram of protons have been accelerated to over 1 GeV in the history of human accelerator physics. This small amount is due in part to the necessary energy requirements. In order to accelerate the 10^{12} protons to an energy of 400 GeV, a power of 100 kWh would be required at 100% efficiency. If a thousand times as many particles were to be accelerated, the power requirements would approach that of a large generating station and the cost would be $100,000 per hour of operation. Furthermore, we shall see that available equipment for detecting and analyzing particles functions well with existing beam intensities.

Another method for accelerating particles avoids large magnetic forces. Instead, the charges move in a straight line down a long tube. An electric force is made to travel down the tube at the same rate as the particles, always acting to accelerate them. This approach is used in what are called linear electron accelerators, which have been used to produce beams of electrons of very high energy. At present, a linear accelerator two miles long is in operation at Stanford, Calif., which can accelerate electrons to about 15 GeV. Linear accelerators are also used to accelerate protons up to about 1 GeV as a preliminary step to their further acceleration in a synchrotron. In recent years, linear accelerators have also been made to accelerate heavy ions, such as an iron nucleus with most of its electrons removed. Although these ions do not reach the same high energies per particle as do protons in a synchrotron, they can be accelerated to a high enough energy to undergo atomic and nuclear interactions, and there is a thriving study of such reactions at several laboratories.

Let us consider the typical process involved in a modern high-energy physics experiment with accelerated particles. The particles, say protons, are first preaccelerated to a convenient energy for injection into the main accelerator. After entering the ring, they remain in a fixed orbit, receiving periodic increases in energy from an electric force that acts on them along sections of the ring. The protons remain in the ring for several seconds, which implies that at a velocity near that of light, they travel for millions of miles, or for many thousands of revolutions. Finally, upon reaching the maximum energy, which is determined by how large the magnetic force that confines them to the ring can be made, the protons are ejected from the ring, usually by the effect of an additional electric or magnetic force. Up to this point, there has been nothing accomplished beyond the transfer of a great deal of electrical energy into kinetic energy of the particles. In order to learn something, these energetic particles must interact with something.

Perhaps the most fundamental choice to be made is whether the experimenter wishes to study the interactions of the particles that have been accelerated, or to use those particles as a method of producing other high-energy particles, whose interactions are the ones of interest to him. This distinction is referred to as a difference between primary and secondary beams. In order to study the interactions and properties of some unstable particles, whose lifetimes are too short for them to be accelerated to high energies, it has been useful to produce secondary beams of them. This is accomplished by allowing accelerated protons to interact with a "target" made of some convenient material. As we shall see, unstable particles called *pions* are often produced in such collisions. Electrically charged pions can be "focussed" through the action of magnetic forces, in a way analogous to the focussing of a light beam by a lens. In this way, a secondary beam of pions can be produced. These secondary beams are neither as energetic nor as intense as the primary beams, but their energy and intensity are sufficient to carry out experiments with the pions. It has also been possible to produce secondary beams of some neutral particles, such as neutrons. Most unstable particles have lifetimes too short even to be used in secondary beams, and these must be studied as they are produced one by one in collisions.

It is much simpler to study the interactions of the particles being accelerated, as they are available in much larger quantities, either inside the accelerator or in an extracted beam. These particles can be allowed to hit a target containing protons, and perhaps neutrons bound in nuclei, and the beam particles will interact with the target particles. If the beam particles have energies of many GeV, the fact that the target particles are bound in a nucleus is relatively unimportant, since this binding energy is much smaller than the kinetic energy. The main qualification to this approach is that after a single collision it is possible for the collision products to interact again with other particles in the same nucleus, producing a change in the results that would be obtained for a single free particle. We can correct for this effect, and obtain the free-particle results from the data. It is also possible to eliminate this problem in the scattering of one proton by another proton, by using hydrogen as a target, in which truly negligible molecular binding is involved.

Before discussing the techniques for detecting the results of collisions, I will mention one disadvantage of the whole procedure of scattering high-energy particles from a fixed target, and some alternatives that have been used. When two particles collide, the results of the scattering such as the cross section might in principle depend separately on the energy of each particle. Actually, it depends only on a combination of the two energies known as the "center-of-mass energy." This is the total energy of the two particles as measured by an observer for whom they have equal and opposite momentum. The reason for this is that according to the relativity principle, the cross section must be the same for any two observers in relative motion. Two such observers will in general measure different values for the energy of each particle, but will obtain the same value for the center-of-mass energy, which is equal to the rest energy of the two particles observed as a unit and is therefore the same for all observers. Thus a dependence on this quantity is allowed by the relativity principle, while a dependence on the individual energies is not.

When the kinetic energy of a beam particle is small compared to the target particles' rest energies, the center-of-mass energy increases proportionately to the energy of the beam particle. Therefore, experiments done at various low beam energies will measure the cross sections at correspondingly different center-of-mass energies and so yield useful information. However, when the beam particles have energy equal to or greater than the target particle rest energy, about 1 GeV for a proton target, the center-of-mass energy begins to vary less than the beam energy. Eventually, for high beam energies, it takes an increase by a factor of 4 in beam energy to double the center-of-mass energy. Because of this, experiments at different high beam energies correspond to not so different center-of-mass energies. For example, if electrons are scattered from electrons in a target, a beam energy of 1 GeV corresponds to a center-of-mass energy of 30 MeV. In order to obtain a collision at a center-of-mass energy of 1 GeV, it would be necessary to give the beam electrons energies of 1,000 GeV, which is quite difficult. Furthermore, the relativity principle also requires that the center-of-mass energy must be greater than the rest energy of any particles created in the collision, so that very high beam energies may be required to create new particles in collisions with stationary targets.

A way to avoid these problems is to arrange for both colliding particles to be moving in opposite directions. In the ideal case, the two momenta are equal and opposite, and the total energy of the two particles is the center-of-mass energy, so that any change in this quantity will probe the cross sections at different values and give new information. In the example given above of a collision between two electrons, it would only be necessary for each electron's energy to be 500 MeV in order to attain a center-of-mass energy of 1 GeV.

This kind of arrangement is known as a colliding beams experiment. It is not practical to do colliding beams experiments by building two distinct accelerators near one another and letting their beams cross. The problem is that the density of particles in the beam is so low that each beam makes a very rarefied target for the other beam, and very few collisions can be expected. In order to get around this, it is necessary to increase the number of particles in each beam by a large factor. This is most easily accomplished by storing the particles accelerated in each cycle of the machine for many cycles, and so gradually increasing the intensity of the beam. This has been done through the use of what is known as a storage ring.

Roughly speaking, this storage ring consists of an extra accelerator ring with a magnetic force but no accelerating electric force. The particles are first accelerated to high energy in a regular accelerator, such as a synchrotron or a linear accelerator. They are extracted from this accelerator by the action of additional magnetic forces and guided into the extra ring, which is carefully designed so that the particles can circulate for very long periods without losing energy or making collisions. Thus the total number of particles in the storage ring can be substantially increased over the number in the extracted beam.

Several different experimental arrangements are possible with storage rings. In one, used with the proton synchrotron at CERN, an acronym in French for the European Center for Nuclear Research, a single ring is used to store two beams of protons traveling in opposite directions. Of interest are the collisions between the protons in the two beams, each of which contains some 10^{14} protons. The beams are allowed to cross each other many times, and interact with a total center-of-mass energy of 56 GeV, equivalent to a collision of a proton of energy 1,500 GeV with a stationary target. This is a higher energy than any available accelerator can produce so that the CERN storage rings can probe aspects of proton–proton scattering that are otherwise inaccessible. However, only proton–proton scattering can be studied in this way. Furthermore, because the total number of interactions is small, it is impossible to produce useful secondary beams from the storage rings. Hence it should not be thought that a storage ring can replace a single ring accelerator of much higher energy for all purposes.

Bubble Chambers and Spark Chambers

Of equal importance to the accelerator is the equipment used to detect the particles that are scattered or created in the collisions. Indeed, this equipment is now comparable in cost to accelerators only a few years old. Several different approaches are used to detect and measure these subatomic particles, and each approach has its characteristic advantages and disadvantages. Hence different experiments make use of one or another of these approaches, depending on the aim of the experiment.

All of the methods for detecting subatomic particles rely, directly or indirectly, on the electrical force between charges. The nuclear forces do not give a sufficient number of events along the path of a particle to make its detection easily possible. This may be seen by comparing the cross section for a typical atomic electromagnetic scattering process to that of a nuclear scattering. It turns out that the large size of atoms compared to the small size of nuclei more than compensates for the weakness of electric force compared to the nuclear force. A charged particle will produce observable effects, such as ionization of the atom, through the electrical interaction of the particle with the electrons in the atom many times in each centimeter of path length. This may be compared to a distance of many centimeters between effects due to collisions with nuclei. These observable effects can be used to define a track for a charged subatomic particle, as well as to measure certain of its properties, such as energy. Conversely, a neutral subatomic particle will not in general be directly detectable, and the detection of these particles relies on their conversion into charged particles, and then using the properties of the latter to infer the properties of the neutral particles.

The two most widely used detection instruments in contemporary subatomic physics research are the *bubble chamber* and the spark chamber. Both of these instruments were developed in the years 1950–1960, and both are used to record the path of a charged particle through space during a well-defined time interval

BUBBLE CHAMBERS AND SPARK CHAMBERS 159

in which the detector operates. Furthermore, each can be hooked up to a photographic or electronic recording system which can store the information about the paths of several subatomic particles that pass through the detector together. This information can then be analyzed subsequently to reconstruct the details of each subatomic event. Finally, thousands or millions of records of different events can be obtained in a reasonable exposure time of the detector to a source of particles, and the results analyzed to give the various probability distributions that characterize the specific process under study.

The operation of both the bubble chamber and the spark chamber are based on a simple idea. An unstable situation is set up in which a large scale change can be triggered by the relatively small amount (on a macroscopic scale) of energy that is lost when a high-energy subatomic particle passes through matter. In the bubble chamber, the situation consists of a liquid that is heated above its boiling point. Such a liquid is called superheated, and can persist without boiling at least for a short time until some stimulus pushes it over the brink and it begins to boil. In the bubble chamber, the stimulus is the entrance of a single high-energy charged particle. As the particle moves through the superheated liquid, it transfers energy to the liquid by ionization of the atoms. This energy is sufficient to begin the process of boiling in the region near the particle, and a small bubble of gas is formed. As the particle progresses through the chamber it continues to lose energy and form bubbles along its path. Since the particle is moving very rapidly compared to the bubbles, for a short time the bubbles lie in a line along the path of the particle. This line of bubbles forms a track that scatters light rays, and so may be photographed by high-speed techniques; the result is a picture indicating that a particle has passed through the bubble chamber. The particle is only observable if its track is at least a millimeter in length. Consequently, only stable particles, or metastable ones—those with half-lives greater than 10^{-11} sec—can be detected in this way. Particles with shorter half-lives decay long before they travel far enough to give a perceptible track.

The bubble chamber is sensitive to the passage of particles for a certain period under 10^{-2} sec following its initial superheating, after which it must be recycled. However, this period is very long

compared to the time any high-energy particle remains in the
chamber anyway, so the bubble chamber typically detects a large
number of particles in each cycle of operation, many of which
have no relevance to the process of interest (Plate 4). As a result,
the analysis of bubble-chamber pictures to find the events of interest is far from a simple operation. In the early days of bubble-
chamber operation, large numbers of people (scanners) were em-
ployed to examine the pictures and locate "events" on them that
interested the experimenters. More recently, there has been progress in developing automated procedures for doing this, so that
the pictures themselves often need not be examined by human
beings. Instead, a computer will analyze millions of pictures and
print out the data about the subatomic events contained in them.
The final step, in which the computer also writes the article
describing the results of the experiment, has not yet been taken.

A spark chamber consists basically of a series of plates that are
alternatively at positive or negative voltage and gaps between
them filled by some gas. When the charge on the plates is high
enough, a spark can jump from one plate to the next, just as in the
discharge tubes discussed in Chap. II. However, this only happens
when the gas between the plates contains some ions. Such ions are
produced in the gas by the passage of a high-energy-charged particle. A spark will then jump the gap between the plates, approximately along the path of the charged particle. As the particle
passes from gap to gap, a series of sparks is built up, which indicates the path of the particle through the whole chamber (Plate
5). This can be photographed, or can be recorded electronically
and reconstructed by computer simulation. One advantage of
spark chambers over bubble chambers is that the recycling time
can be made much shorter, as only charged particles, rather than
macroscopic bubbles, are in motion. As a result, it is possible to
get "cleaner" pictures in a spark chamber, in which only the event
of interest is seen, as in Plate 5. On the other hand, a disadvantage of spark chambers is that they tend to be insensitive to
particles that are traveling nearly parallel to the plates because
sparks are not easily formed in this direction, whereas bubble
chambers are equally good at detecting particles moving in any direction. This is not a serious problem for particles in the incident
beam, as the spark chamber can be positioned so that these parti-

cles will all move perpendicular to the plates. However, these chambers are often used to detect particles created within the chamber, and since these can emerge in any direction, some particles can be lost.

A further difference between the two detectors is that bubble chambers tend to be general purpose detectors, which can be used for many different types of experiments, hence they tend to be built and left in operation for long periods. On the other hand, spark chambers are usually modified for each individual experiment, in order to make use of their very fast recycling time. Therefore, permanent spark chambers are rare, although individual ones can be used for several different experiments.

Each of the detectors can be used in several ways, although some are most common to each. A bubble chamber is used simultaneously as a target and analyzer. This means that the incident particle beam interacts with the atoms in the bubble-chamber liquid, undergoing various collision processes. The charged particles produced in these collisions are then detected in the bubble chamber, their trajectories measured, etc. By varying the liquid in the bubble chamber, it is possible to measure the interaction of a given incident beam with various targets, such as protons, deuterons, or heavier nuclei. It is sometimes desirable to use high-density liquids such as freon in the bubble chamber in order to maximize the amount of material available for interactions with the primary beam. Also, such dense liquids increase the probability of detection of photons through the photon's creation of electrons when it passes near a nucleus. A bubble-chamber picture showing the first discovery of a particle is given in Plate 6. The discovery was made by Nicholas Samios and his co-workers at Brookhaven Laboratory in Upton, N.Y.

The spark chamber can also be used as a combined target and analyzer. In doing so, it has an added advantage that more material can be included in the metal plates, thus increasing the total number of atoms in the target. The first demonstration that two types of neutrino exist came in a spark-chamber experiment of this kind, in which the target material had a total mass of ten tons, corresponding to about 10^{31} nucleons. Even with such a large target, only about 50 neutrino-induced events were seen. Spark chambers are also effective in detecting photons, or neutral particles that decay into photons, because the photons convert into charged particles in the plates.

It is possible to use spark chambers to analyze particles produced outside the chamber. This is useful if one wishes to obtain information about the scattering of some particle by free protons alone, since in the spark chamber both protons and neutrons are present bound in nuclei. If the beam particles are scattered in a hydrogen target, and the products allowed to pass through a spark chamber, their properties can be measured, and information obtained about the original scattering by protons. This is also possible in a hydrogen-filled bubble chamber, but the analysis would be more complicated because of the complexity of each picture.

The Analysis of Particles in Detectors

Thus far, we have discussed the detection of charged particles qualitatively, but for most experiments it is necessary not only to detect the particles, but also to identify them and to measure their properties. To a large extent, it is possible to do this through their interactions with the material in the detector, or with electromagnetic forces that are made to act on them while they are in the detector. The identification of particles is made somewhat easier by the fact that the particles of each charge come with widely separated masses, so that a determination of mass and charge is usually enough to identify the particles. The determination of the mass of a particle can be done in several ways. One is to measure its energy and momentum and then use the relation between energy, momentum, and mass that is implied by relativity theory.

The momentum of a charged particle is most easily measured by observing its motion under the action of a constant magnetic force. In this circumstance, the particle will move in a circular arc, whose curvature varies inversely with the momentum of the particle. This curvature can be measured from the track of the particle in a bubble chamber, for example, and the momentum determined. This method depends on the assumption that the magnitude of the electric charge of the particle is known, which is usually the case because all known subatomic particles with measurable tracks have the same magnitude of charge. The direction of curvature of the path under a known magnetic force also determines the sign of the electric charge, since this direction is opposite for particles of opposite charge.

The energy of a charged particle can be measured, under many conditions, by determining the rate at which this energy is lost, by ionization, to the medium through which the particle moves. Perhaps the simplest way of measuring this is in the case in which the particle loses enough energy by ionization that it comes to rest while still in the detector. In that case, the total distance that it moves, known as the range, depends only on its initial energy and on the medium, so that a measurement of the range determines the energy of the particle. This technique works very well if the particle energy is not so high that it cannot come to rest. For particles with energies of a GeV or more, the range in liquid hydrogen is usually much larger than the size of the detector and such particles do not come to rest. Furthermore, for such particles, the amount of ionization produced in moving through equal distances is the same for all particles of equal charge. These facts make the identification of high-energy particles rather difficult, and special techniques to do this must be invented for each experiment. This becomes a severe problem if many charged particles are created in a collision, because there may be no easy way of deciding which of several types of charged particle a given track may be. The analysis of such events to obtain a detailed description of the particles produced is often impossible, and the events are usually classified just by the number of tracks detected.

One distinction that can often be made even for high-energy particles is between electrons and other particles. This can be done by exploiting the special properties of electrons. Because of their low mass, electrons tend to emit photons easily when they collide with atoms, especially in materials with high nuclear charge. This gives electrons an additional way of losing energy, which is often more important than ionization. Furthermore, the photons emitted can, by a process described in Chap. VIII, create more electrons, which then emit more photons, etc. This whole process leads to a large number of electrons and photons traveling in approximately the same direction and is known as a "shower." The formation of such showers is usually an indication that an electron, or sometimes a photon, has been produced in a collision or a decay. It is sometimes possible to measure the energy of this primary electron or photon by adding up all the energies of the particles in the shower, as the latter energies are eventually low enough for the particles to stop or be absorbed. Other subatomic particles have much greater mass than electrons and do not produce showers.

The techniques for the acceleration and detection of particles described above have all become routine to physicists. Nevertheless, they are all recent discoveries that were almost unthinkable twenty years ago. It is worth remembering that the progress in physics in this century owes as much to advances in the technology of accelerator design and of particle detectors, as to the abstract ideas of the theoretical physicists.

VIII

Subatomic Particles and Their Interactions

Almost all of the matter ordinarily present in the Universe—on Earth, in the stars, and in space—is made up of three types of particles: electrons, protons, and neutrons. Under the conditions on Earth, these are bound into atoms. However, in the stars, neutral atoms are relatively rare, and the particles are instead found in the form of separated nuclei and electrons known as a plasma. This is because the temperatures in stars are so high that collisions rapidly remove the outer electrons from the atoms, although some of the more tightly bound inner electrons may remain attached to the nuclei. We have seen that the properties of atoms, including the nuclei, can be fairly well understood in terms of the somewhat simpler properties of their constituent particles. This is also true for the properties of the plasma form of matter found in the stars. In view of this, it might be thought that electrons, protons, and neutrons are indeed the fundamental building blocks of matter sought throughout the history of physics.

While this view was prevalent among physicists in the 1930s, we now know that these three particles are but a few among a great number of similar subatomic particles, and that no primary role among these particles is to be associated with the three composing ordinary matter. The study of these subatomic particles has been the central theme of physics in the second half of the twentieth century. While a good deal of knowledge and understanding about these particles has been obtained, many questions remain unanswered. No fundamental new intellectual structures beyond quantum mechanics have been developed, or as yet seem to be needed, to describe the subatomic particles. However, several remarkable characteristics of particle behavior have been seen to follow from the union of quantum mechanics with relativity, most notably the possibility of creating and annihilating particles. Particle physics, or high-energy physics as it is sometimes called, has proceeded through a mixture of experiment and theory, in which now one and now the other has led to new insights. Lately, theory has probably played a greater role, and in many cases, particle physics might be described by the paraphrase, "theorists propose and experiments dispose." Whether this pattern will continue in the future development of particle physics is difficult to predict.

The Creation and Annihilation of Particles

Because most of the subatomic particles that we now know are not found in ordinary matter, it has been necessary to perform elaborate experiments to produce them and to study their properties. In a typical particle physics experiment, a beam of protons that have been accelerated to very high energy is made to impinge on a piece of matter at rest. The protons in the beam collide with protons and neutrons in the nuclei of the atoms in the matter. Under these conditions, the binding energies of the nucleons are unimportant, and the collisions take place as if a proton were hitting an unbound neutron or proton. This is analogous to the situation in which two blocks are held together by a thin string. If one block is hit lightly, the system of two blocks and string will move together. However, if the block is given a very strong blow, the force acting on the block is greater than the maximum resistance of the string, and the block struck will fly away, breaking the string and leaving the other block behind. So a nucleon in a nucleus that is hit by a high-energy proton will usually be ejected from the nucleus.

If the target hit by a beam of protons is about a meter long, essentially all of the protons in the beam will encounter at least one nucleus in passing through the target, and interact with the nucleus in some way. One possible result of this is that the proton may be scattered through some angle, as was described previously for low-energy electrons and photons. However, for high-energy particles, the characteristic phenomenon is instead the production of new particles, not present in the initial beam or in the target matter. In such a production process, some of the kinetic energy of the initial particles is converted into the rest energy of the extra particles that are produced. This is a confirmation of one of the possibilities predicted by relativity theory. The production of extra particles occurs in more than 50% of high-energy proton collisions. The number of extra particles produced in each collision increases as the energy increases, beginning at one or two when the incident proton has a kinetic energy of 1,000 MeV (1 GeV). These extra particles arc usually not protons or neutrons.

Particles with such kinetic energies of 1,000 or more MeV are not very common in the Universe. While sometimes they occur in what is called cosmic radiation, a stream of high-energy particles that seems to be present throughout the Universe and which is known to continuously hit the Earth, the number of particles present in this cosmic radiation is less than 10^{-8} of the total number of particles in the Universe at least if estimates based on the cosmic radiation reaching Earth are accurate. The only other place that high-energy particles are known to occur is in physics laboratories. Actually, the number of particles accelerated to high energies in physics laboratories is approaching the total number reaching Earth is cosmic radiation.

Before discussing the nature and properties of the other subatomic particles, it is worth indicating why physicists are convinced that they are *produced* in the collision, rather than being liberated. We have seen that electrons and neutrons are also found by probing matter in various ways and observing what emerges, but in those cases, we believe that the particles emerging were already present, rather than being produced by the probes. We must, therefore, establish some criterion to determine which of these al-

ternatives is occurring. One of the criteria for determining whether
a particle is already present in matter is whether it is possible to
liberate it by a variety of different projectiles. The other criterion
physicists use is whether the extra particle emerges from collisions
in which the total kinetic energy of the initial particles is smaller
than the rest energy of the extra particles. According to the conservation of energy, the extra particles can be *created* only if there
is enough kinetic energy to be changed into their rest energy.
When applied to various experiments this criterion leads to the
conclusion that electrons, protons, and neutrons are already present in matter because they can be liberated from it in collisions
involving particles of kinetic energy much lower than any of the
rest energies. Since this does not happen for any of the other particles, we conclude that they are not present in ordinary matter.
However, these criteria become somewhat ambiguous in circumstances in which the binding energy of a particle is comparable to
its rest energy, because the binding energy must always be sup-
plied through kinetic energy of the initial particles.

The particles that have been produced in collisions have a wide
variety of masses, spins, and other properties. In some respects,
the complexity of the phenomena of subatomic particles is comparable to that of atoms or nuclei. A handicap to studying their
properties in addition to the difficulty of their production is that
most of these particles decay spontaneously into others in an extremely short time after they are produced.

Of the particles not found in ordinary matter the easiest to produce in collisions, is the positron or positive electron discovered in
1934 by Carl Anderson. If a beam of photons, or of electrons,
whose energy is above a few MeV is passed through thin metal
plates, there are found emerging from the plates electrons and
other charged particles whose mass is the same as that of electrons, but whose charge is positive and equal to that of the proton
rather than that of the electron. Careful measurements show that
these positrons are produced together with an additional electron,
or "pair produced" as it is called (see Plate 4). As an estimate of
the magnitude of this effect, a photon of 5 MeV passing through
10 cm of iron is likely to make one electron and one positron,
known as a pair. Production of positrons by photons is only observed to take place when the energy of the photon is at least 1.02
MeV, equal to the sum of the rest energies of an electron and

positron. This indicates that the positrons are not ejected from the matter by the photon, in which case the energy necessary to do this would probably vary with the material, but rather are created by the photons, which disappear in the process of creation. In the latter case, the conservation of charge would require that a negative particle be created along with the positive particle, as is observed. The conservation of energy also requires that the photon have at least the total rest energy of the pair that is created. This energy is low enough that some gamma rays from radioactive nuclear decays can produce pairs, and so positrons can be produced without particle accelerators.

Further evidence for the creation hypothesis comes from the observation of the inverse process of pair annihilation in which a positron, moving through matter, collides with an electron, and the two convert into two photons whose total energy is equal to 1.02 MeV, or somewhat more if the positron is moving rapidly. Since positrons and electrons are seen to "disappear" into photons in this way, it is less implausible that photons can materialize into massive particles as described. The annihilation process also helps to account for the absence of positrons in ordinary matter. Since this matter contains large numbers of electrons, any positrons that happen to wander by will annihilate rapidly, usually in 10^{-8} sec or less in solid materials. This does not imply that electrons are somehow more fundamental than positrons, because they are more abundant any more than the greater abundance of iron than gold on Earth implies that iron is a more fundamental material than gold. The reason for the huge surplus of electrons over positrons, at least in our part of the Universe, is unknown. It has been suggested that in other parts of the Universe the situation is reversed; there, positrons together with negative protons, or antiprotons, which have also been discovered, predominate. However, no evidence for this conjecture exists.

An interesting illustration of the use of the principle of relativity is to understand why a photon in vacuum cannot produce an electron–positron pair, or why a pair in a vacuum cannot annihilate into a single photon. If these processes were possible, they would have to be possible for all observers in uniform relative motion, according to the relativity principle. Suppose then that an electron and positron could annihilate to produce a single photon. For any electron and positron, there exists an observer moving with some velocity such that the total momentum of the electron and positron vanishes. The total energy will not be zero, because it is at least equal to the sum of the rest energies, or 1.02 MeV. By the laws of conservation of momentum and energy, the photon seen by this observer must also have zero momentum and nonzero energy. But this is impossible for a photon, since its energy and momentum are proportional. Hence the process cannot occur for this observer, and, by the relativity principle, therefore cannot occur for any observer. No such conclusion can be reached for annihilation into two photons, as it is quite possible for two photons going in opposite directions to have zero total momentum but not have zero energy.

Although electron–positron pairs usually produce photons when they annihilate, if their kinetic energy is high enough they will sometimes produce other subatomic particles. The study of these alternative annihilation modes has produced interesting data about subatomic particles.

Several very general features of particle physics are illustrated by the observations involving electrons and positrons. One is that different particles can freely convert into one another, provided that the conservation laws, such as energy, momentum, charge, and certain more recently discovered ones, are satisfied. From this point of view, the photon must be thought of as a particle similar to the others, but with somewhat different properties, such as zero mass and no charge. While this conclusion about the impermanence of the subatomic particles was already suggested by phenomena like beta decay in which electrons are created, it was most forcefully brought to the attention of physicists by the phenomena involving positrons. In the forty years since the discovery of electron–positron pair production, the conclusion has been confirmed again and again through the discovery of many other types of particles that can be created or annihilated, sometimes singly and sometimes in pairs. These creation or annihilation phenomena are closely related to another phenomenon displayed by some of the subatomic particles, that of spontaneous decay of one particle into several less massive particles. We have discussed one example of such a decay, that of the free neutron into a proton, electron, and neutrino. In creations, annihilations, or decays there is a transformation of one or more particles into other particles, consistent with the over-all conservation laws. Of course, not every combination of particles will change into any other combination, and one of the main problems of subatomic particle physics has been the determination of which transformations occur and why.

Another general property of particles suggested by the existence of positrons is the occurrence of an *antiparticle* associated with each type of particle. An antiparticle of a particle is defined as a particle with the same mass and spin, but with opposite electric charge and opposite value for certain other properties similar to charge. Thus the positron is the electron's antiparticle. The photon, having no charge at all, would be indistinguishable from its antiparticle, and so is identical to it. All of the known electrically charged particles, as well as many of the electrically neutral ones such as the neutron, have known antiparticles whose properties other than charge agree with those of the particle. This result is no accident, but rather a general result following from any theory that satisfies both quantum mechanics and the relativity principle. The existence of the positron, the first antiparticle to be discovered, was actually predicted by Dirac several years before its discovery on the basis of such a theory. The existence of the antiproton and antineutron, the antiparticles of the proton and neutron, were confidently predicted on a similar basis, and an expensive accelerator was built with the expectation of producing antiprotons. This actually was achieved in 1955, thus avoiding many red faces among physicists, as well as the need for embarrassing explanations to Congress, which provided the funds for the accelerator.

The relation between particle and antiparticle is symmetric, in that if *A* is the antiparticle of *B,* then *B* is equally well the antiparticle of *A.* The antiparticles of the electron, proton, and neutron are much less common in our part of the Universe that these particles, and therefore the antiparticles seem more mysterious. For the other subatomic particles that have been discovered, both particles and antiparticles are uncommon, and it is partly a matter of convention which we call the particle and which the antiparticle. From a theoretical point of view, since the two types occur symmetrically in a theory neither is more fundamental than the other. The symmetry between particles and their antiparticles is so complete that it is possible to build up complex forms of matter out of the antiparticles to the particles of ordinary matter. Thus antihydrogen atoms, consisting of a negatively charged antiproton and a positron, can exist, with properties similar to those of ordinary hydrogen atoms. Similarly, antideuterons, or nuclei consisting of an antiproton and an antineutron have been produced with properties similar to deuterons.

On the other hand, matter containing both a particle and its antiparticle would be very unstable, and would annihilate rapidly. Thus atoms known as positronium, consisting of an electron and positron bound together by their mutual electric attraction are sometimes produced briefly when positrons travel through matter. These positronium atoms are somewhat similar to ordinary atoms, having a lowest-energy level with a series of excited levels, and radiative transitions between levels have been observed. However, in a time varying between 10^{-7} and 10^{-10} sec, the electron and positron annihilate each other, producing two or more photons, and the positronium atom disappears. Thus bulk matter made of positronium is impossible. Atoms consisting of a proton and antiproton annihilate even more rapidly, and likewise cannot occur in bulk. An important difference here is that protons and antiprotons generally annihilate into several other lighter particles, called pions, rather than into photons. This is an indication of a fact we shall consider later: protons can interact in other ways than electrons. Also, protons can annihilate with antineutrons, and neutrons with antiprotons, as well as with their own antiparticles, in each case producing several pions. This implies that annihilation is something more than a cancellation of a particle with its exact opposite, as the name might suggest. Instead, both annihilation of a particle and antiparticle, and the pair creation of particle and antiparticle are qualitatively similar to other transformations that can take place among subatomic particles.

Pions

The production of particle–antiparticle pairs is not the dominant process in collisions of high-energy nucleons. Nucleons rarely produce electron–positron pairs in collisions, because to do so they must first produce photons, which is a relatively improbable occurrence. While nucleons with more than a few GeV of kinetic energy do sometimes produce nucleon–antinucleon pairs, this is also relatively uncommon. Instead, the three kinds of particles that are most commonly produced in collisions are called pi-mesons or *pions*. The three types of pions have electric charges of $+e$, $-e$, and zero; in other words equal to those of the proton, electron, and neutron. The positive and negative pion are each other's antiparticle. The neutral pion, like the photon, is identical to its antiparticle. The masses of the two charged pions are equal and about 270 times that of the electron, corresponding to a rest energy of about 140 MeV. The uncharged, or neutral, pion has a slightly lower mass than the charged ones. All of them have zero spin and all are bosons. The three pions are produced in approximately equal numbers in collisions of high-energy protons with matter, although in an individual collision there may be a surplus of one type. The average number of pions produced in a collision between two protons is about 1 when the proton's kinetic energy is 1 GeV, and increases slowly with increasing kinetic energy to about 10 at 1,000 GeV.

Since these pions are produced so easily in high-energy collisions, one may wonder why the world is not as filled with them as it is with electrons, protons, and neutrons. The answer is that the pions are all unstable, and transform, in a very short time, into particles with lower mass. As a result, any pion that is made in a collision exists for only a short time. The neutral pions live only about 10^{-16} sec after being produced, and therefore can travel only about 10^{-5} cm before decaying. It is therefore not possible to make a beam of neutral pions and their properties must be studied by more indirect methods. A neutral pion usually *decays* into two photons, similar to nuclear gamma rays but of higher energy, averaging 70 MeV each. For the two charged pions, the half-life is approximately 10^{-8} sec if the pion is moving slowly, and somewhat longer if it is moving rapidly, because of the time-expanding effect of rapid motion that is predicted by relativity theory. This short period of time is nevertheless long enough to study the charged pions, with the techniques of modern electronics. It is even long enough for a charged pion moving at close to the speed of light to travel for many meters before it decays. As a result, it has been possible to make beams of charged pions that have been produced by collisions and to study their interaction with protons and neutrons by passing these beams through matter.

The charged pions do not decay directly into particles found in ordinary matter. Instead, they usually decay into another charged particle about 200 times heavier than an electron called a muon together with a neutral particle, whose mass is zero, or at least very small, called a mu-neutrino (Plate 7). For a while it was believed that the mu-neutrino was the same massless neutral particle that appears in nuclear beta decay. However, ingenious experiments first done in 1962 showed that the particles are not identical, and the particle which appears together with an electron in beta decay, is now called the "e-neutrino," to distinguish it from the one that appears with muons.

The muons, which occur as particle and antiparticle with opposite signs of charge, are also fermions like electrons and carry a spin of ½ $\hbar$. Like pions, muons are also unstable. They have a half-life of 10^{-6} sec and decay into electrons or positrons along with two neutral particles believed, but not yet proven to be, one of each type of neutrino (see Plate 7). We have seen in Chap. VI that muons can form atoms with ordinary nuclei, which last for a short time, until the muon decays, or is absorbed.

The reasoning behind the conclusion that charged pion decays involve only one neutral particle while muon decays involve two such particles is of interest, since it is quite difficult in either case to detect the neutral particles directly. However, one can measure the energy of the charged particles by deflecting them in magnetic forces or by measuring the rate at which they cause ionization in gases. When this is done, it is found that the muons that emerge from charged pion decays have a unique kinetic energy, depending only on the the energy of the decaying pion. If the pion decays at rest, which happens if it loses its kinetic energy by passing through matter, the muon energy is about 4 MeV. On the other hand, the electrons that emerge from muons that decay at rest do not have a unique kinetic energy, but instead have a range of energies from zero up to about 52 MeV. What this means is that if many muons are observed to decay, and the electron energy is measured in each case, it will be found that the kinetic energy of the electrons varies from case to case, but always lies in the range from zero to 52 MeV. The precise value of the electron's energy for an individual decay is given randomly, just as the time of decay of the muon is. This is another example of how subatomic systems allow only statistical predictions of their behavior.

These results can be easily understood on the basis of the conservation of energy and momentum, provided that a single neutral particle emerges with the muon while two emerge with the electron. Suppose a particle at rest decays into two particles of equal mass. The two emerging particles must have equal but opposite momentum in order that the total momentum add to zero, and be conserved. But then the kinetic energies of the two particles must also be equal, since the momentum of a particle determines its kinetic energy. Therefore, each particle will have a unique kinetic energy given by one-half of the difference between the rest energy of the decaying particle and the two rest energies of the emerging particles. A similar conclusion holds if the emerging particles have different masses, although in this case the kinetic energies of the emerging particles will be unequal. The basic idea is that the two conservation laws of energy and momentum fix the kinetic energies of the two emerging particles.

On the other hand, if three or more particles emerge, there are various ways in which they can share the momentum, still adding to zero. For example, two particles could go in one direction, each with one-half unit of momentum, while the third goes in the opposite direction with a full unit of momentum. The three particles can also go in different directions, as long as the total momentum adds to zero along all directions. Under these circumstances, the conservation of energy will not fix the energy of each particle, but instead these energies will depend on the way they have shared the momentum. Therefore, any one of the particles can have a range of energies allowed by the conservation laws. These conclusions can be partly checked in other decays, in which three particles, all of which can be observed, are produced, and it is found that although the individual energies and momenta vary, the total energy and momentum is conserved. We saw in Chap. VI that arguments similar to this were originally applied to nuclear beta decay, in which the electrons emerge with a range of energies, to conclude that an unobserved neutral particle, the e-neutrino, must be produced with the electron.

It is in some ways fortunate that charged pions decay into muons, because it is much more difficult to produce muons directly than to produce pions. If pions did not decay into muons, the muon would not have been discovered so easily. While it is possible to produce pairs of muons by photons—just as happens for electrons—much-higher-energy photons are needed because of the higher muon rest energy. Furthermore, even when the photon energy is high enough, the rate of producing muon pairs is some 40,000 times less than the rate of producing electron pairs, so the process is hard to detect. What is more significant is that muons, although less massive than pions, are very rarely produced directly in the collisions between high-energy protons that produce pions so copiously. If a proton of 10,000 MeV hits another proton it will usually make several pions, whereas it will produce a muon directly less than once in 10,000 such collisions. Of course, muons will be seen to emerge from the region in which the collision occurs, but these muons will almost all be the result of production of pions or of certain other particles which then decay into muons. Careful measurements of the energy and momentum of the emerging particles can distinguish such secondary production of muons from direct production in the collision, and the conclusion is that the latter rarely occurs. It is also possible in some circumstances to "see" the chain of events occur with a bubble chamber (see Plate 7). This clearly shows that the production of the pion is the primary process, and that the muon emerges in a later decay of the pion.

Interactions

The type of transformation among subatomic particles exemplified by pion production or decay or electron–positron pair creation is described by physicists through a new idea known as an *interaction*. Interactions are a generalization of the idea of force or potential energy in Newtonian mechanics, in the sense that a force is a very special kind of interaction. The need for this new concept comes from two aspects of subatomic physics. One is that since motion in a well-defined orbit must be abandoned because of Heisenberg's relation, a force in the sense of a definite change of motion is a less useful notion than in Newtonian mechanics. The changes in motion of an object in quantum mechanics are more discontinuous than the Newtonian picture. The other aspect is the creation and annihilation of particles which cannot be described by forces at all.

Instead, the concept of an interaction is used to describe any change in the state of a system described by quantum mechanics. The attributes of the state may include not only the quantum numbers we have discussed previously, but also the numbers of different types of particles that are present. For example, a state may contain an electron and a positron, each with some momentum and spin projection. This state may change, in time, into another state, still containing an electron and positron, but with different momenta because a scattering has taken place. Alternatively, it might change into a state with two photons because of pair annihilation. In either case, we would say that some interaction has occurred, and the strength of the interaction would be measured by the probability of the transition from one state to the other.

In the simple description just given, two independent interactions would be used: one to describe the scattering, the other to describe the annihilation. However, we shall see that in many cases it is possible to relate different interactions to one another by using an important idea of quantum mechanics called virtual transitions. Indeed, one great advantage of the concept of interactions, as compared to Newtonian forces, is that all the phenomena we know involving atoms and their subatomic constituents can be described in terms of a small number of distinct interactions. In particular, many of the different forces used in Newtonian physics such as cohesive forces, chemical attractions, etc., are found to be different aspects of the same fundamental interaction.

An interaction, then, is anything that can change the state of a system existing at one time into another state at another time. We cannot, in the present stage of physics, "explain" interactions in terms of anything more fundamental, any more than forces could be "explained" in Newtonian physics. Instead, what we have done is to realize that subatomic particles can change into one another under certain restrictions, and the notion of an interaction is a description within quantum mechanics of the rules that we have found such changes to obey.

The rarity of direct production of muons, compared to pions, is an important clue to a major feature of the interactions of subatomic particles—the varying strengths with which those interactions occur. In particular, muons interact much less with other particles than do pions or nucleons. Another indication of this is given by the behavior of a beam of muons passing through matter. Although muons have a half-life of 10^{-6} sec, they can travel 100 meters (m) or more before decaying, and therefore intense beams of muons can easily be made. If such a beam of muons, with an energy of about 1,000 MeV for each muon, is passed through matter, very few of the muons will be scattered or otherwise affected by the matter unless the beam passes through many meters of material. Muons of higher energy such as are sometimes found in cosmic radiation may pass through miles of earth without significant scattering or absorption. This is to be contrasted with a beam of pions or protons, which will be scattered or absorbed significantly by a 1-m layer of material. Furthermore, the scatterings that the muons do undergo will be generally elastic, i.e., no other particles will be produced, and involve just the electrical interaction between the muon and the atoms in the material. On the other hand, pions or protons of high energy will often produce new particles when they pass through matter and many aspects of their behavior are not describable by electrical interactions, or even by the nuclear force that we know exists between the protons in the beam and the particles in the nucleus.

The different phenomena displayed by muons and by pions, both in their production and in their passage through matter, has led physicists to conclude that pions have an additional way of interacting with matter than do muons. Because this other mechanism leads to relatively large production of pions and to relatively large scattering of pions, it is generally called a strong interaction, as distinct from the electric and magnetic interactions we have considered and the *weak interaction* involved in decays of pions and muons which we shall consider later. With some exceptions, the known particles all have strong interactions with one another. One exception is the photon, which interacts only with charged particles. The other exceptions include the muon, electron, their partners the mu-neutrino and e-neutrino, and the antiparticles of all of them. This latter group all have spin $\frac{1}{2} \hbar$, and are collectively called *leptons*.

TABLE I

Leptons

PARTICLE NAME AND SYMBOL·	REST ENERGY	HALF-LIFE.	ELECTRIC CHARGES
e-neutrino and e-antineutrino $(\nu_e, \overline{\nu_e})$	0	Stable	0
Mu-neutrino and mu-antineutrino $(\nu_\mu, \overline{\nu_\mu})$	0	Stable	0
Electron and positron (e^-, e^+)	0.51 MeV	Stable	$-e, +e$
Negative and positive muon (μ^-, μ^+)	105.6 MeV	1.5×10^{-6} sec	$-e, +e$

The known leptons. All have spin $\frac{1}{2} \hbar$.

From the way in which the strong pion interaction was discovered, it is clear this process must involve at least pions and protons or neutrons. Furthermore, since pions can be created as well as scattered by this mechanism, it must be more complex than a simple force between objects that remain unchanged. In fact, we shall see that the nuclear force is itself a secondary manifestation of the strong interaction of pions with nucleons. In order to allow for the emission and the absorption of pions, the strong pion interaction must involve a change in the number of pions when it occurs. That is, it must involve a transition from a state containing some number of pions to another state containing more or less pions. It can be seen that if this basic interaction always involves a change by one in the number of pions, it is possible then to account for all other changes involving pions by combinations of this basic interaction. For example, if we want to describe a process in which a neutral pion scatters from a proton, we can think of this as occurring in two steps. The original pion may be absorbed by the proton leaving a proton, after which the proton may emit the pion that emerges. While it is also possible to imagine an interaction process that scatters the pion in a single step, such a process could not by itself lead to the creation or destruction of a single pion, and so an additional mechanism would be needed to account for these processes. By taking the single pion creation or destruction as the primary process, we also can account for scattering.

A problem whose solution indicates the need for a quantum-mechanical description of these processes arises when we do this. Imagine a single neutral pion passing near a proton. The strong pion–proton interaction will supposedly lead, with some probability, to the absorption of the pion by the proton, producing a proton in motion. However, it can be shown that this process cannot occur according to the laws of conservation of energy and momentum. To see this, let us again invoke the relativity principle, and describe the process as it would be seen by an observer for whom the pion and proton originally have equal but opposite momentum. The total momentum of this two-body system is thus zero. After the pion is absorbed, the momentum of the proton that remains must also be zero. However, this means that the proton will have no kinetic energy either. But originally the pion and proton did have kinetic energy, and the pion also had its rest energy which disappeared when it was absorbed. The conclusion is that the energy of the original pion–proton system was greater than the energy of the remaining proton, in contradiction to the law of conservation of energy. By using the relativity principle it can be shown that the initial and final energies can never balance no matter what the total initial momentum of the system is.

This implies that if the conservation laws are strictly satisfied, the basic strong interaction of the emission or absorption of a pion by a nucleon cannot take place. However, this conclusion is stronger than the analysis warrants. We have already seen in our discussion of quantum mechanics that quantities such as the energy and momentum of atoms and subatomic particles are subject to the indeterminacies of Heisenberg's relation. Consequently, the conservation laws concerning these quantities must be interpreted somewhat differently than they are in Newtonian physics where no such indeterminacy exists. In particular, the law of conservation of energy must be interpreted as meaning that for an experiment in which the energy of some system is measured extremely accurately before and after a change occurs in the system, the two values of the energy will agree within the limits of accuracy of the measurement. The Heisenberg relation, however, implies that very accurate energy measurements can be done only if a great deal of time is available for the measurement—a direct analogy to the statement that accurate measurement of momentum can be done only if the system is free to occupy a large region of space. Because of this restriction on the circumstances in which energy can be measured accurately, it follows that the energy conservation law is strictly valid only if the energy is compared at times that are sufficiently far apart. If the energy is measured at two times very close to one another, the inaccuracies inherent in the measurement will be large, and the measurements may disagree within the wide limits of these uncertainties.

The numerical values of the relation are such that a time of 10^{-23} sec requires an energy uncertainty of about 100 MeV, or about that of the pion's rest energy. This time interval is about what it takes for a pion moving at the speed of light to move a distance corresponding to the average separation of two nuclei in a nucleus. For a time as long (!) as 10^{-10} sec, the energy uncertainty drops to 10^{-5} eV, and is thus less than the accuracy of standard laboratory measurements. Therefore this sort of loophole in the law of conservation of energy is really only relevant to processes that occur on a time scale pertinent to atomic and subatomic processes, and there is no contradiction between it and the high accuracy with which the law is known to hold in macroscopic physics. We have already learned in Chap. IV that this reformulation in the energy conservation law allows the occurrence, for a short time, of virtual transitions in processes such as the scattering of light by atoms which would be very different without them. The effect of virtual transitions, in which extra particles are created and destroyed for a short time, on processes involving subatomic particles is even more important. All of the properties of these particles, either alone or in combinations, are influenced by the possibility of transitions that are done and then undone within a very short time interval. These transitions can involve the creation and annihilation of massive particles such as pions, or even nucleon–antinucleon pairs, without contradicting the conservation of energy as long as the creation and the anihilation both occur within a time interval that is less than $\hbar$ divided by the amount by which energy changes in the transition. For the creation of a single pion by a proton at rest, this minimum energy change is about 140 MeV, corresponding to a maximum time of 10^{-23} sec. For creation of an antiproton–proton pair, the time is even shorter.

It might be thought that because the particles created in virtual transitions exist for such a short time, they could have little effect on the properties of the particles that remain permanently. This is not the case. Consider a system consisting of a neutron and proton bound together, the deuteron. The proton can emit a positive pion and convert to a neutron. This pion can, within the time it is allowed to exist, travel to the neutron and be absorbed there within the time allowed. Since the pion that is emitted and absorbed carries energy and momentum with it, the effect of this is to transfer energy and momentum between the proton and neutron (Fig. 18). But such a transfer of energy and momentum is what is otherwise known as a force between the two particles, since a force is just our name for the observation that when objects are near each other their individual momenta and energy can change. Therefore, one consequence of the creation and annihilation of pions by nucleons is the existence of a force between the nucleons. The properties of this force are qualitatively in agreement with those of the nuclear force discussed in Chap. VII, although more complex considerations determine whether the force is attractive or repulsive. One example of a property predicted by this pion exchange model is that if the neutron and proton are farther apart than the pion can move in the time that it may virtually exist there should be no force between them. This distance is about 10^{-13} cm, in agreement with the observed range of the nuclear force. Furthermore, similar forces will occur between two neutrons, between two protons, or between a neutron and proton, because it is possible to exchange neutral pions as well as charged pions. Since all pions have approximately the same mass, the range of the forces will all be about the same. The strength of the force depends on how often the nucleons emit and absorb pions, that is, on the strength of the pion–nucleon interaction. In this

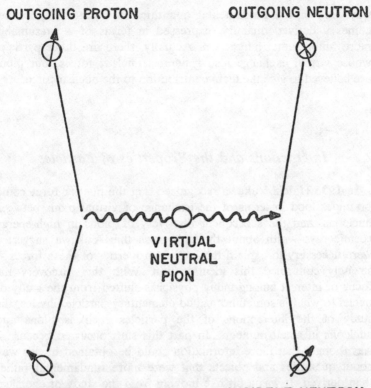

OUTGOING PROTON   OUTGOING NEUTRON

VIRTUAL
NEUTRAL
PION

INCIDENT PROTON   INCIDENT NEUTRON

FIGURE 18. Nuclear force arising from virtual pion exchange. The nuclear force is believed to arise from a two-step process involving virtual pions. Here one way is shown in which the force between a neutron and proton arises. The proton, indicated by an open circle with an arrow, emits a virtual neutral pion, indicated by an open circle, and the proton's momentum changes. The pion travels a short distance to a nearby neutron, indicated by a circle with a cross and an arrow. There it is absorbed, changing the neutron's momentum. The two-step process results in an exchange of momentum between neutron and proton, which is what we call a force between the two particles. Diagrams such as this are widely used by physicists as a tool for remembering the various virtual processes that can generate some observed process. They are called Feynman diagrams, after Richard Feynman, the physicist who first used them.

way, one of the fundamental quantities in nuclear physics becomes a derived quantity, expressed in terms of a presumably more fundamental interaction. Actually, there are other particles whose virtual exchange also generates nuclear forces, but pions are believed to give the main contribution to the nuclear force.

Interactions and the Properties of Particles

In 1935 Hideki Yukawa recognized that the nuclear force could be understood in terms of the exchange of virtual pions between nucleons, and the subsequent discovery of pions in high-energy cosmic rays—with about the properties that Yukawa suggested were necessary to explain the known properties of these forces—strongly confirmed this identification. With this discovery the focus of interest among many physicists shifted from the study of nuclei to what is sometimes called elementary particle physics, the study of the interactions of the particles such as pions and nucleons in small numbers. In part this shift occurred because it was thought that more information could be obtained in this way about quantities and objects that were more fundamental rather than derived. It was a bit like moving from the study of chemical compounds to the study of the elements composing those compounds. In each of these cases, the results have been somewhat mixed insofar as understanding the complex systems is concerned, although very much important information about their components has been obtained.

This reduction of the nuclear force to successive pion emission and absorption is an example of how the idea of force as a fundamental description of the way particles influence each other has been replaced by the idea of an interaction, involving creation or annihilation of particles, rather than simply a change of momentum. This revision of thought has even occurred for the well-known electric and magnetic forces which can now be pictured as the result of virtual emission of a photon by one charged body, followed by its absorption by another charged body. The emission and absorption of photons by charges is called the *electromagnetic interaction.* In this way the forces are related to other known properties of photons and charges without the need to assume an independent mechanism through which two charges influence one another. In that respect, the explanation of forces through virtual emission and absorption represents a decrease in the number of independent hypotheses, and so is philosophically appropriate. As far as we now know, it is reasonable to think of all the forces in nature in these terms, although in the case of gravitational forces we have not yet observed the particles (gravitons) that would be required to transmit the force.

There are other ways in which virtual creation processes affect properties of particles. Consider a single neutron. It has been found that the neutron generates a magnetic force associated with its spin, just as does the proton and electron. Since the neutron has no total charge, its magnetic force is somewhat surprising, and indicates that the neutron, to some extent, consists of positive and negative moving charges whose total charge adds to zero, but which generate the magnetic force. One way of understanding this result is to consider that the neutron can make a virtual transition into a proton and a negative pion. Since these particles are charged, they can produce magnetic forces. Therefore, in this model, the neutron will produce magnetic forces proportional to the probability of its virtual transitions into a proton–pion state. It is irrelevant to this model of the neutron that the virtual transitions occur over a very short time period. This only implies, as we have seen, that the virtual pions cannot travel more than about 10^{-13} cm from where they are produced. But the usual measurements of magnetic forces detect moving charges anywhere near the center of the neutron, no matter how close. In order to demonstrate that these charges are actually restricted to within a radius of 10^{-13} cm or less it is necesary to measure how the magnetic force is distributed in space. This can be done, even over such short distances, by electron scattering from neutrons similar to that discussed for nuclei. These measurements show that the charges that produce the magnetic force of a neutron and which also make the magnetic force of a proton deviate from the value expected for a structureless spinning charge, are indeed confined to within a radius of less than 10^{-13} cm of the center of the distribution. This result is therefore consistent with the idea that virtual transitions are responsible for this magnetic force. It has been more difficult to obtain a precise numerical value for the force from this model primarily because the many subatomic particles now known that can occur in virtual transitions can also contribute to the magnetic forces of nucleons. Therefore one must regard this model as not yet completely demonstrated.

The idea that the charge and mass of a proton or neutron are not located at a point, but instead are distributed over a region with a size of 10^{-13} cm or so implies that these nucleons are not simple particles, but instead have a structure, just as atoms and nuclei were previously found to have. The structure of nucleons arising from virtual transitions is of a different sort than that which would arise from a simple composite picture of the nucleons, although the later type of structure may also exist, as we shall see in Chap. IX.

Like any other transitions described by quantum mechanics, the virtual transitions do not occur at fixed intervals, but only according to certain probability laws. In any brief interval of time, as allowed by the Heisenberg's relation, a proton may be found to have converted by the strong pion–nucleon interaction into a neutron and pion, with a certain probability. This will also be true in any other time interval of equal length. So we may think of the proton as being in a constant over-all state that is a combination of a state with a single proton, a state with a neutron and pion, or other states containing any of the other subatomic particles into which a proton can make a virtual transition. Although these other states do not have the same energy as a proton, they do have the same charge. While it has not been possible to obtain precise estimates of the relative amounts of these various configurations in a proton, it appears that configurations such as a neutron and pion are approximately as probable as the single proton configuration itself. This is another measure of the strength of the pion–nucleon interaction, and the large probability of the neutron–pion configuration is another reason for referring to this interaction as strong. This complicated combination of states has the properties we associate with a proton as measured in the laboratory, rather than the ideal situation of an isolated proton without any of the objects into which it can make virtual transitions.

This is sometimes described by calling the protons found in nature "dressed" protons, as opposed to the "bare" protons without their cloak of virtually created particles. Thus the possibility of creation of particles, together with the concept of virtual transitions that is inherent in Heisenberg's relation, leads to a wholly new notion of the complexities of what originally seemed a simple system—an isolated subatomic particle.

One of the more important consequences of this discovery is that the properties of the various subatomic particles are interconnected in an elaborate way not previously suspected. As an example, we have seen that the magnetic force produced by a neutron may arise from virtual production of charged pions. If charged pions did not exist, or if they had very different masses or spins than they do, it is likely that the neutron's magnetic force would have a quite different value. Conversely, the same strong pion–nucleon interaction that allows for virtual creation of pions by a nucleon also allows for virtual transformations of a pion into a nucleon–antinucleon pair. These transformations influence the properties of pions. For example, it is thought that the decay of the neutral pion into two photons occurs through a multistep process in which the pion changes virtually into a proton–antiproton pair, after which this pair annihilates into photons, much as an electron–positron pair does. A fairly accurate numerical value for the lifetime of the neutral pion can be obtained from this model.

Arguments of this type have often been used either to predict new particles or to predict some of the unknown properties of known particles. For example, the very existence of the neutral pion was suspected in advance of its discovery because of the known equality of the nuclear forces between two protons and that between neutron and proton. Only neutral pions can generate a proton–proton force by the one-pion exchange mechanism we have considered, and in order for this force to agree in magnitude with the neutron–proton force which arises at least in part from charged pion exchange, the neutral pions must have about the same mass as the charged pions and the strength of the interaction of each type of pion with nucleons must be simply related. All of those expectations have been confirmed by experiment, and are an indication of what has been an important theme in particle physics: the relations, or symmetries, among the properties and interactions of different particles.

The expression of such relationships is most easily done through the language of a branch of mathematics called group theory, but some aspects of it can be seen without any elaborate mathematics. The basic idea is that when particles can interact with each other and so influence each other's properties through virtual creation and annihilation, any simple relationship that holds between the properties of one group of particles must be mirrored in corresponding relationships among the properties of particles in the other interacting groups. For example, the neutron and proton have approximately equal masses, e.g., the difference between their masses is $\frac{1}{10}\%$ of the mass of either. The masses of the proton and neutron are influenced by virtual transitions in which these particles emit and then reabsorb both neutral and charged pions. One way to see this is to recognize that a dressed proton is partly a state containing a bare proton and a neutral pion. The mass of the dressed proton, then, is a sum of terms corresponding to this state and the other states it may be. A similar result holds for the dressed neutron. Suppose now that the probability of virtual emission of a neutral pion by a proton were different than it is for a neutron. Then that part of the total neutron mass coming from the neutron–neutral pion component of the state would be different from the corresponding part for the proton coming from the proton–neutral pion state. Numerical estimates suggest that these contributions may be a substantial part of the mass, so that if they were unequal it would be very surprising that the total masses of the neutron and proton are as close as they are. The inescapable conclusion is that the neutral pion interacts in a very similar way with the neutron and proton. We have already seen that the approximate equality of the neutron–neutron, neutron–proton, and proton–proton nuclear forces implies a similar equality between the interactions of charged and neutral pions with nucleons. Thus by observing symmetries in the properties of nucleons, and assuming that to some extent these properties originate from their interactions with pions, we are led to symmetries in the pion–nucleon interaction and in the properties of pions. As other particles have been discovered that interact strongly with neutrons and protons, the same arguments have been used to place restrictions on the properties that such particles must have. The relationship between particle properties inferred from these considerations is known as charge independence, because it relates the masses and interactions of particles with different electric charges like neutron and proton. It is also sometimes called *isospin symmetry*, because there exist similarities between the mathematical description of this symmetry and of the intrinsic spin of particles.

For a number of years in the 1940s it was believed that pions, nucleons, and their antiparticles were the only strongly interacting particles, and that a suitable mathematical description of their interactions would lead to a complete understanding of all the phenomena involving these particles, both in isolation and bound in nuclei. This view gradually became untenable as new particles were discovered that also interacted strongly with pions and nucleons and with each other. The work in subatomic physics since about 1950 has largely been devoted to uncovering and understanding these new particles and their interrelations. We shall consider this work in Chap. IX.

IX

Hadrons and Quarks

Of the subatomic particles discovered in the period from 1940 to 1975, the vast majority fall into the family of those that interact strongly with one another—the *hadrons*. The hadrons include all the known particles, except for the photon, electron, muon, two neutrinos, and their antiparticles. Most of the hadrons are unstable, and decay, slowly or rapidly, into the few known stable particles. The precise number of known hadrons depends somewhat on the criterion used to distinguish two particles. If we regard as distinct any two particles with different values of at least one of the quantities, rest mass and electric charge, then there are many hundred known hadrons. On the other hand, if we identify as different aspects of the same particle such objects as the charged and neutral pion, or the proton and neutron, and other particles that are now grouped into subfamilies, the number is much smaller. In any case, the hadrons clearly form a complex collection of objects, comparable in its amount of detail to that of the chemical elements.

The Classification of Hadrons

In the presence of this surplus of hadrons, it was important to develop classification schemes to relate the particles to one another. Important progress in this direction was made in the 1960s through a classification known as *unitary symmetry,* a mathematical generalization of the isospin symmetry mentioned previously. Unitary symmetry, like isospin symmetry, relates the properties of particles which have the same spin but different masses and charges. Other, complementary classification schemes have been used to relate particles which have the same charge but different mass and different spin. The latter schemes, which will not be discussed in detail, are somewhat analogous to the classification of states of individual atoms or nuclei, in which excited states with different spin than the ground state are quite common. Most of the known hadrons can be grouped into families through the use of one of these types of classifications in a way reminiscent of the grouping of chemical elements by Mendeleev in the nineteenth century. Just as the success of Mendeleev's Periodic Table led eventually to the question of what structural properties of atoms implied the relationships of the Periodic Table, so the success of unitary symmetry and other classification schemes for subatomic particles leads to the question of whether these particles might themselves have some inner structure that is reflected in the classification schemes. This question has received a partly affirmative answer through the so-called *quark* model of hadrons which we will discuss. But let us first consider some of the properties of hadrons that this model attempts to explain.

Within the hadrons there are several subclasses whose properties play an important role in understanding the behavior of hadrons. The major distinction is between two types of particles known as *baryons* and mesons. Originally, the distinction was made on the basis of rest mass. The first known mesons, the pions, had much lower rest mass than the first known baryons, the nucleons. However, the two types are now known to contain particles with a wide range of masses, and the heaviest mesons are comparable in mass with the heaviest known baryons. Nevertheless, an important distinction exists among the hadrons, which serves to define baryons and mesons. The baryons all carry a conserved quantity that is not carried by the mesons. A conserved quantity is simply one whose value, added together for all the particles participating in a reaction, does not change from the beginning to the end of the reaction. Electric charge is one example of a conserved quantity. The evidence that some hadrons carry a conserved quantity distinct from electric charge can be seen by examining certain transformations that are observed to occur among hadrons, and others that are not observed to occur.

Consider first a reaction involving any combination of nucleons (either unbound, or bound in nuclei), antinucleons, and pions. Suppose we count each nucleon as carrying $+1$ unit of some quantity, each antinucleon as carrying -1 unit, and pions as carrying 0 units. It is found empirically that if we add up the values of this quantity for all the particles in a reaction, the total value at the beginning will equal the total value at the end, for any reaction that is observed to occur. The property that is carried by nucleons and antinucleons, but not by pions is called baryon number, and the result that its total value never changes in a reaction is called the law of conservation of baryon number.

As an example, consider a reaction in which an antiproton hits a deuteron. At the beginning there are a proton, neutron, and antiproton present, so that the total baryon number is $(+1)+(+1)+(-1)$, or $+1$. Therefore, the particles emerging from the reaction must also carry a total baryon number of $+1$. These particles might consist of a neutron and some pions, or two neutrons and an antineutron, or many other combinations of particles, all with a total baryon number of $+1$. It would be inconsistent with the baryon number conservation law for an antineutron and pions to be the only particles emerging, as these would have a total baryon number of -1. In fact, the latter reaction is not observed. In a similar way, the conservation of electric charge forbids certain reactions from occurring. For example, it would forbid a single proton from being produced in the reaction we are considering. However, since the neutron carries baryon number but not electric charge, it should be clear that different quantities are involved in these two laws.

When we consider reactions involving particles other than nucleons, antinucleons, and pions, the assignment of baryon number can be extended in a way that allows the conservation law to remain valid. Leptons and photons are assigned baryon number 0. For the other "simple" hadrons, i.e., those that are not atoms or nuclei, the baryon number is $+1$ for particles with half-integer spin, -1 for antiparticles with half-integer spin, and zero for all integer spin hadrons. For combinations of hadrons, such as complex nuclei, the baryon number is just the sum of the baryon numbers of the constituents. In making these assignments, the decision as to whether a baryon is a particle or an antiparticle is made so that the baryon conservation law is satisfied with that assignment. It is an empirical result that such an assignment is always possible.

In summary, the precise definition of baryons is those hadrons that have nonzero baryon number, while the mesons are hadrons with zero baryon number. All of the mesons therefore have integer spin, while all the simple baryons, i.e., those with baryon number 1, have half-integer spin. The usual practice among physicists is to use the term baryon mainly for those hadrons with baryon number 1 and I shall follow that rule. The association of spin with baryon number among the hadrons is at this stage somewhat fortuitous. In principle, a hadron could exist with integer spin and baryon number $+1$. For example, if a hadron were discovered that decays into a neutron and an electron, it would be assigned a baryon number of $+1$, and would have integer spin. Such hadrons have not been found, and some explanation of this fact is given in the quark model discussed below.

The baryon conservation law forbids many processes that might otherwise occur, such as a transformation of two protons into two positive pions. Since all particles with less rest energy than a proton have baryon number zero, the combination of the baryon number conservation law and the energy conservation law leads to the conclusion that the proton should be absolutely stable. It also implies that for any baryon number other than zero there will be a lowest-energy state which will be absolutely stable, i.e., cannot decay into lower-mass particles. Those conclusions are both in agreement with observations. For example, by examining a large sample of matter for possible decay of the protons it contains into mesons or leptons, it has been found that the half-life for proton decay cannot be less than 10^{28} years, and there is no reason to believe that isolated protons ever decay. Of course, the protons in some unstable nuclei can decay into neutrons, because the total energy of the proton in the nucleus is more than that of the neutron; this cannot happen for an isolated proton, for which there is no lower-energy state containing a single baryon. The very high limit for the lifetime of the proton implies that all known interactions—

strong, electromagnetic, and weak—must conserve the number of baryons. If there are any interactions that do not conserve the baryon number, they must be much weaker than any of the known interactions, even than gravitational interactions. Baryon conservation also implies that in the decay of all unstable baryons, such as the Λ^0 (lambda-zero) particle, a baryon must appear among the decay products, so that all such decays eventually lead to a proton, perhaps after a chain of decays. For example, a particle known as the Ω^- (omega-minus), decays into another particle known as Ξ^0 (Xi-zero) and a negative pion. The Ξ^0 then decays into Λ^0, and a neutral pion. Finally, the Λ^0 decays into proton and negative pion. In each of these decays, the baryon number is preserved, and the Λ^0, Ξ^0, and Ω^- each can be assigned a baryon number of $+1$ (Plate 6).

Of course, baryon conservation does allow such reactions as the annihilation of a proton with an antineutron into several pions, since the total baryon number at the beginning and end of this reaction are both zero. Therefore annihilation reactions involving baryons and antibaryons of low kinetic energy can produce particles in the final state with much more kinetic energy than reactions involving only low-kinetic-energy baryons can. In the former reactions, the rest energy of the baryon and antibaryon that annihilate gets transformed into kinetic energy of particles produced, while in the latter type of reactions, the rest energy of the baryons must remain in this form in the final state, as the baryons cannot disappear. This means that in the absence of a large supply of antibaryons, the rest energy of the nucleons that compose ordinary matter is essentially unavailable to us, and the most energetically favorable reactions we can find are those in which the nucleons are made to bind together more tightly in nuclei which release at most 1% of the total rest energy. It is useless as an energy source to produce antibaryons and then make them annihilate with nucleons as it would take at least as much energy to produce the antibaryons as one could obtain from their annihilation.

The hadrons are further subdivided according to the value of another quantity that some of them carry, known as *hypercharge*. Just as with the baryon number, the hypercharge of a particle can be a positive integer, negative integer, or zero, multiple of some unit value, and the hypercharge of a system of particles is the sum of the values for the individual particles. The hypercharge of a particle and its antiparticle are opposite in sign. Therefore, those particles with nonzero hypercharge are distinct from their antiparticles. Hypercharge differs from baryon number in two important ways. One is that both some baryons and some mesons have nonzero hypercharge. The other is that while hypercharge is conserved in strong interactions and electromagnetic interactions, there exists a third class of interactions, called *weak interactions,* in which hypercharge is not conserved. It is therefore possible under some circumstances for the total hypercharge of a system to change. This possibility has an important implication: whereas the lowest-mass particle with nonzero baryon number, the proton, is absolutely stable, the lowest-mass particles with hypercharge, which are two particles called charged and neutral K mesons, are only metastable. That is, while the K mesons cannot decay into less massive hadrons such as pions through strong or electromagnetic interactions, they can and do decay into pions, and also into leptons, by weak interactions. As suggested by the name weak interactions, the rate of these decays is relatively small, and the corresponding half-lives of the particles are in the range of 10^{-7} to 10^{-10} sec, rather than the 10^{-15} to 10^{-19} for particles that decay by electromagnetic interactions, or 10^{-21} to 10^{-24} for particles that decay by strong interactions. This has the effect that particles decaying by weak interactions often travel sufficiently far between their production and their decay that they will show a measurable track in a detecting instrument, if they are electrically charged.

This would not be the case for particles that decay by strong interactions. For such a particle, with a typical half-life of 10^{-22} sec, the distance it can travel, even at the speed of light, before decaying is only 3×10^{-12} cm, or much less than one atomic diameter, which is too small a motion to detect at present. Indeed, these particles are detectable only through their decay products, and the lifetimes of such particles are measured indirectly by using a result related to Heisenberg's relation for energy and time.

According to this result, if a particle has a half-life T, its energy cannot be precisely determined, but instead will be uncertain by an amount ΔE, where the product of ΔE and T is approximately equal to $\hbar$. When the decay products of the particle are observed, their total energy will vary over a range whose size is ΔE. That is, if many decays of the same type of particle are observed, the total energy of the products will vary from case to case within a range of values whose maximum and minimum differ by ΔE. This quantity ΔE is called the decay width of the decaying particle. For particles that decay by strong interactions, the width is usually 1 MeV or more, and can be readily measured. The half-life of the particle can then be determined by using Heisenberg's relation. Particles that decay by weak interactions also have decay widths, but these are much smaller, since the lifetimes are much longer, and these widths cannot be measured directly at present. Instead, the half-lives of such particles are measured directly, by observing how far the particle travels between production and decay. Particles that decay by electromagnetic interactions tend to fall between these two methods, and their half-lives have been measured by complex indirect methods.

In practice, physicists tend to classify as metastable any subatomic particle that does not decay by strong interactions. By this criterion, one stable and seven metastable particles with baryon number $+1$, and with spin $\frac{1}{2}\, \hbar$ are known.

We have already met the stable proton and metastable neutron which both have hypercharge $+1$. Somewhat higher in rest energy than these is a baryon of hypercharge 0, known as Λ^0. This particle must decay by weak interactions for the following reasons. According to baryon number conservation, a baryon must be present in the decay products of Λ^0. The only lower-mass baryons are the nucleons, and these have hypercharge $+1$. Therefore if the Λ^0 decayed by strong or electromagnetic interactions, the conservation of hypercharge would require that a meson with baryon number 0 and hypercharge -1 be present among the decay products also. The particles with the lowest rest energy having these properties are the K mesons. However, the rest energy of Λ^0 is less than the sum of the rest energies of the nucleon and K meson, so that energy conservation forbids such decays. Therefore, the Λ^0 does not decay by strong or electromagnetic interactions, and instead decays by weak interactions, usually into a nucleon and a pion.

Slightly higher in energy than Λ^0 are three metastable baryons with hypercharge 0 known as Σ^+ (sigma-plus), Σ^- (sigma-minus), and Σ^0 (sigma-zero). The related symbols for these parti-

TABLE II

Stable and Metastable Hadrons

PARTICLE NAME AND SYMBOL	REST ENERGY in MeV	Spin	BARYON Number	Hypercharge
Mesons				
Neutral pion (π^0)	135	0	0	0
Positive and negative pion (π^+, π^-)	140	0	0	0
Positive and negative K-meson (K^+, K^-)	494	0	0	+1, −1
Neutral K-mesons ($K^0, \overline{K^0}$)	498	0	0	+1, −1
Eta meson (η^0)	549	0	0	0
Baryons				
Proton and antiproton ($p, \bar{p}$)	938	$\frac{1}{2}\hbar$	+1, −1	+1, −1
Neutron and antineutron ($n, \bar{n}$)	940	$\frac{1}{2}\hbar$	+1, −1	+1, −1
Lambda and antilambda ($\Lambda^0, \overline{\Lambda^0}$)	1116	$\frac{1}{2}\hbar$	+1, −1	0, 0
Sigma-plus and anti-sigma-plus ($\Sigma^+, \overline{\Sigma^+}$)	1189	$\frac{1}{2}\hbar$	+1, −1	0, 0
Sigma-zero and anti-sigma-zero ($\Sigma^0, \overline{\Sigma^0}$)	1192	$\frac{1}{2}\hbar$	+1, −1	0, 0
Sigma-minus and anti-sigma-minus ($\Sigma^-, \overline{\Sigma^-}$)	1197	$\frac{1}{2}\hbar$	+1, −1	0, 0
Xi-zero and anti-xi-zero ($\Xi^0, \overline{\Xi^0}$)	1315	$\frac{1}{2}\hbar$	+1, −1	−1, +1
Xi-minus and anti-xi-minus ($\Xi^-, \overline{\Xi^-}$)	1321	$\frac{1}{2}\hbar$	+1, −1	−1, +1
Omega-minus and anti-omega-minus ($\Omega^-, \overline{\Omega^-}$)	1672	$\frac{3}{2}\hbar$	+1, −1	−2, +2

The hadrons that are stable or metastable against decay by strong interactions. Particles and antiparticles are listed together. The pattern of rest energies, spins, baryon numbers, and hypercharges are well accounted for by the quark model of these and the other unstable hadrons.

cles are to indicate that they form an isospin *multiplet*, like the three pions. The Σ^+ and Σ^- cannot decay by strong or electromagnetic interactions for reasons similar to Λ^0, and thus decay by weak interactions. The Σ^0 also cannot decay by strong interactions, but does decay by electromagnetic interactions into Λ^0 and a photon. This is possible because the zero rest energy of the photon means that it can be emitted with very low total energy, and so the small difference in rest energy between Σ^0 and Λ^0 is sufficient to allow this decay, whereas the difference in energy is not enough to allow the emission of a pion by strong interaction.

Finally, there are two metastable baryons with still higher rest energy and hypercharge —1, the Ξ⁻ (xi-minus) and Ξ⁰ (xi-zero), which can decay into the hypercharge zero Λ^0 together with a pion by weak interactions. There is also a metastable baryon of hypercharge —2, the Ω^-, which has spin $3/2\,\hbar$ and decays by weak interactions into a Ξ particle and a pion.

From these results, it becomes clear that a particle will decay primarily by the strongest interaction that it is allowed to use by the conservation laws. If a particle can decay by strong interactions, it will do so. If this is forbidden by conservation laws, it will decay by electromagnetic interactions, which are $1/100$ as strong. If this is also forbidden, the particle will decay by weak interactions, which are much weaker still. A particle which decays by strong interactions may nevertheless sometimes decay by electromagnetic interactions, as all these processes can occur independently. However, it is usually difficult to detect any but the primary form of decay because of the great disparity of strengths among the interactions.

There are eight metastable particles known among the mesons. These include the three pions, the two K mesons and their two antiparticles, and the η^0 (eta-zero) particle, which decays via electromagnetic interactions into pions and photons.

No other stable or metastable hadrons were known before 1976, and it is thought that if more exist, they must have some new qualities, akin to hypercharge, which would keep them from decaying by strong interactions. One such possibility, the "charmed" particles, will be discussed below.

While the metastable baryons and mesons cannot decay by strong interactions, they can be produced by strong interactions. However, they must be produced in combinations of two or more particles, such that hypercharge, baryon number, and electric charge are conserved. For example, in a collision between a pion and a proton, the total baryon number and total hypercharge are both +1. If a K^+ meson with hypercharge +1 and baryon number 0 is produced, it must be together with one or more particles with hypercharge zero and baryon number 1, such as a Λ^0. This type of "associated production" is in fact what has been observed.

When it was first observed that some particles were easy to produce, but decayed slowly, it was thought strange, and these particles such as K mesons were called strange particles. Even though this phenomenon has been explained by the conservation of hypercharge, the term strange particle is still sometimes used.

The relatively long lifetimes of the metastable hadrons implies that they can travel observable distances between their production point and their decay point. It is therefore possible to observe either tracks or gaps due to these particles in detecting instruments. This, together with the relatively low rest energy of the metastable hadrons is why they were detected before many of the other hadrons. There are many other examples of hadrons with the same hypercharge and baryon number as some of the metastable hadrons, but with higher rest energies. These heavier hadrons can and do decay into the lighter hadrons by strong interactions, and therefore have lifetimes so short that they cannot be directly observed. These very unstable hadrons are often called resonances, or excited states, in analogy to the excited states of atoms. Most of the known hadrons are of this type. It is useful to think of these very unstable hadrons as involving the same constituents as the metastable hadrons, with some rearrangement of the internal motions of these constituents. Such a picture is indeed given in the quark model. In any case, all of the hadrons are equally significant for understanding the phenomena of subatomic particles. The proton is not more significant than an unstable hadron, any more than the ground state of an atom is more significant for understanding the atom than any of the excited states.

We can summarize the results of studying the pattern of decays of the hadrons by the conclusion that each hadron may be assigned a value of three different properties—electric charge, baryon number, and hypercharge. The first two of these are conserved in all known interactions, the last is not conserved in weak interactions. An important requirement of any theory of hadrons is to account for these conservation laws.

The classification of hadrons has some aspects that cut across the subdivisions made according to the three conserved properties. We have mentioned the idea of isospin symmetry, which identifies as three aspects of one particle the three types of pion which have different electric charge. This can be done because certain properties of all three pions, their mass, spin, and strong interactions, are approximately or exactly the same. As an extension of this idea, Murray Gell-Mann proposed the idea of unitary symmetry, in which the eight stable or metastable baryons of spin $\frac{1}{2} \hbar$ are identified as aspects of one particle, and the eight metastable mesons are taken to be aspects of another particle. Also, the strong interactions of each group of eight, or octet, are assumed to be related to one another in a mathematically simple way. The particles that are considered as aspects of one particle in this way are referred to as members of a multiplet. This type of identification is analogous to identifying men and women as different aspects of one species, homo sapiens. In order for any such identification to be useful, it is necessary that a significant set of properties of the objects being identified should be similar. For example, the spins of the eight metastable mesons must all be the same, and they are indeed all zero. It is not important that the electric charges of these mesons are not all the same, because the classifying principle refers to the strong interactions which do not depend upon electric charge. A more serious problem is that in some cases, especially for unitary symmetry, the rest energies of particles that are identified are quite different. For example, those of pions and K mesons differ by a factor of 3.5. This implies that some effects exist involving hadrons that do not satisfy unitary symmetry. It is not yet clear what this symmetry-breaking effect is. The breakdown of unitary symmetry with respect to hadron rest energies has variously been attributed to an unknown strong interaction that does not share this symmetry, to the weak interactions, which are known not to satisfy it, or to a difference in the rest energies of the hypothetical quark constituents of hadrons.

Apart from their rest energies, the hadrons obey the dictates of unitary symmetry, which includes isospin symmetry as a special case, fairly accurately. This can be tested by comparing various decay rates of hadrons. Unitary symmetry makes a number of predictions concerning the decays by strong interactions of some of the very unstable hadrons into the metastable hadrons. These predictions have been found to be satisfied to an accuracy of about 10%. For the special case of those relations implied by isospin symmetry, the accuracy of the predictions is even greater, often as good as 1%. The validity of such predictions, together with the eventual discoveries of all of the members of various multiplets, which were not all known when unitary symmetry was first proposed, form the convincing evidence for the applicability of unitary symmetry to the hadrons. A theory about the constitution of the hadrons must therefore account for the validity of this symmetry.

Hadron Scattering Reactions

Although a great deal can be learned about hadrons by studying them singly, still more information is obtained when two or more are near enough to each other to have a mutual influence. This occurs naturally in nuclei, but only for some hadrons. An important role in hadron physics has therefore been played by experiments in which hadrons are brought together artificially in scattering reactions. In such reactions, the energy of the hadrons can be controlled and their properties studied under a wider variety of conditions than in nuclei. Also, some of the metastable hadrons can be used as projectiles in scattering experiments, and information about their properties obtained in that way. A scattering reaction involves an experiment in which two of the stable or metastable hadrons collide, and two or more hadrons emerge. In the simplest such reactions, the same two hadrons that collide also emerge and no new particles are produced. Such collisions are known as elastic, because there is no change in the energy of either particle when viewed by an observer for whom the particles have equal but opposite momentum. These elastic collisions are only a small fraction of the total number of high-energy collisions. The usual results are an inelastic collision. In such collisions, either two particles that are not the same as the original ones emerge, or else more than two particles, which may include the original ones, emerge. In an inelastic collision, the energy of each outgoing particle is different from the energies of the incoming particles, although the total energy of all the particles is of course the same. When two given particles, such as two protons collide, the collision may be either elastic or inelastic (Fig. 19). Of the many possible outcomes to any collision that are allowed by the conservation laws, the one which will occur in a specific case is not precisely determined, and only the relative probabilities of different outcomes is predicted by quantum mechanics.

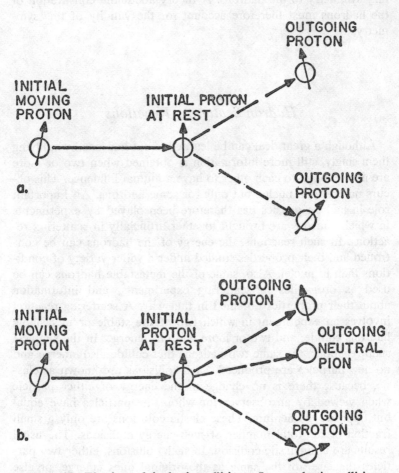

FIGURE 19. Elastic and inelastic collisions. In an elastic collision, shown between two protons in a., no new particles are produced, and only the momentum and spin, if any, of the particles involved changes. In an inelastic collision, in addition to a change in motion, one or more new particles emerges. In b., an inelastic collision between two protons is shown, in which a neutral pion is produced in addition to the protons.

Inelastic reactions can be analyzed with respect to the various particles that emerge. An important point to remember about this analysis is that only the stable or metastable hadrons are observed directly, as the very unstable ones decay before reaching the detector. However, it is possible to determine indirectly when very unstable hadrons have been produced. For example, if an unstable hadron with a certain rest energy is produced and then decays into a pion and a proton, the sum of the pion and proton energies, evaluated by an observer for which their total momentum is zero, will equal the rest energy of the unstable hadron. Because the very unstable hadrons have decay widths, this rest energy is not a precisely defined number, but will vary from event to event. By analyzing in this way many events in which a pion and proton emerge, it is found that the pion energy plus the proton energy gives approximately the same number each time, within the accuracy defined by the decay width. This indicates that a specific unstable particle has been produced which decays into the pion and proton. Both the rest energy and the decay width of the particle can be found from this analysis. Many of the very unstable hadrons were discovered by this type of analysis. It has been found that a substantial fraction of all inelastic scatterings occur via such production of very unstable hadrons.

Since the scattering of hadrons occurs by strong interactions, such scatterings should also satisfy the predictions of the isospin symmetry satisfied by these interactions. This has been verified in a number of cases, although it is more difficult to do so for scattering than for decays because some of the required reactions involve unavailable projectiles such as neutral pions. One example of a test of isospin symmetry in scattering is to compare the scattering of positive and negative pions from deuterons. We have seen that positive and negative pions are related by isospin symmetry. The deuteron is unaffected by isospin symmetry because it contains a combination of a neutron and a proton that the action of the symmetry leaves untouched. In this respect, the deuteron is analogous to an atom with spin zero, which is unaffected by rotational symmetry. The prediction of isospin symmetry is that the cross sections for the scattering of positive and of negative pions by deuterons should be equal, insofar as strong interaction effects are concerned. This prediction is accurately satisfied by the data.

The small difference between the cross sections that is observed can be attributed to the electrical interaction between the charges of the pions and that of the deuterons. This electromagnetic interaction occurs together with the strong interaction and does not satisfy the requirements of isospin invariance, so the combined effect of the two, which is what is observed, does not satisfy this requirement precisely either. However, since the strong interactions are much larger than the electromagnetic interactions, the deviations from isospin symmetry are small. It is believed that all observed deviations from isospin symmetry, such as the small difference in rest energy between neutron and proton, can be attributed to electromagnetic or to weak interactions.

Another interesting property of hadron scattering is the dependence on the energy of the colliding particles of the total cross section for all possible processes. At energies below a few GeV this total cross section varies considerably as the energy changes. However, at higher energies the total cross section becomes approximately constant, or varies very slightly with energy. For example, the total scattering cross section of one proton by another proton varies by less than 10% over an energy range from 5 to 100 GeV. Its value in that range of energies correspond to an effective area for the proton of about 2×10^{-26} cm^2, a value that is not implausible based on a model of strong interactions involving virtual exchange of pions and other hadrons. The total cross sections of other hadron scattering processes have similar values, and are similarly constant over wide ranges of energy. Detailed mathematical models have been invented and applied to these scattering phenomena in order to reproduce these results. Efforts have also been made to explain further details of the scattering such as the relative probabilities of the production of different hadron states in a specific collision and the variation of the production rate with the direction of the outgoing particles. These efforts have been partially successful, but are too mathematical to discuss in this book. Perhaps eventually a simple picture of the apparently complex phenomena of hadron scattering will emerge. More probably, the situation will remain similar to atomic or nuclear scattering which is still difficult to explain in detail, even though the underlying theory is well understood.

The Quark Model of Hadrons

The system of subatomic particles has become so complex in terms of both the number of objects and of their properties, that physicists, who are always searching for simplicity, have become convinced that the observed particles are not the ultimate level of reality, and that something simpler must exist in terms of which the observed particles and their behavior can more easily be understood. This simplification, if accomplished, would be a step in the same direction as understanding the complex properties of different kinds of atoms in terms of the simpler properties of electrons. The most straightforward approach to simplification is the idea that many of the observed particles are composite, made of simpler things. For a while it was thought that some of the known particles might be made of other known particles, such as pions out of nucleons and antinucleons. However, all of the known hadrons seem about equally complex in their behavior, and so that idea did not succeed. Instead, it is now believed that the constituents of the observed particles are other particles that have not yet been observed. Until now (1976), this idea has been seriously applied only to hadrons, since it is those that have been discovered in large numbers. The leptons remain few and relatively simple in their properties, and at the present stage of physics there seems little reason to think about constituents for leptons.

A composite theory of hadrons can be approached in several ways, both theoretical and experimental. Direct experimental evidence of constituents of hadrons has not been found, so the properties of these hypothetical constituents must be inferred from the properties of the hadrons themselves, just as some of the properties of neutrons were inferred from studies of nuclei before free neutrons were discovered. In the most successful model of hadron constituents, these constituents are taken to be new kinds of particles commonly called quarks. The word quark was adapted for obscure reasons from a line in the book *Finnegans Wake*. This was done by Murray Gell-Mann, one of the inventors of the quark model. The model was independently invented by George Zweig. In order to make inferences about the properties of quarks from those of hadrons, we must make some assumptions about the laws that govern the behavior of quarks. It is plausible to assume that quarks are described by the same general principles as the observed hadrons, i.e., by the principles of quantum mechanics and of special relativity. There is no guarantee that this is true, but physicists tend to extrapolate the domain of validity of the laws they know until evidence to the contrary is found.

The first things that we can infer about quarks is the number and types of quarks that may exist in order to account for the observed array of hadrons. We know that the hadrons include baryons with half-integer spin. If only quarks with integer spin existed, then by the rules for combining spins and orbital angular momenta, any composite system of quarks would also have integer angular momentum, and could not represent baryons. Therefore, at least one type of quark must have half-integer spin, and baryons must contain some of this type of quark. The simplest assumption is that all types of quarks have spin $\frac{1}{2}\,\hbar$ and are fermions. In this model, the baryons must contain an odd number of quarks, since by adding an odd number of half-integers we again get a half-integer. The mesons would contain an even number of quarks (or of antiquarks, which would also have spin $\frac{1}{2}\,\hbar$), since adding together an even number of half-integer spins gives an integer spin.

Since we imagine that quarks are a different kind of particle than those previously discovered, it is plausible that a single quark cannot decay into hadrons or leptons, even if is allowed by the conservation of energy. That is, we assume that quarks of any type carry a positive value of some quantity, and that the conservation of this quantity forbids such decays. Antiquarks would carry a negative value for this quantity, which we can call "quark number." We can then account for the conservation of baryon number as follows. Assume that a baryon contains more quarks than antiquarks, while a meson contains equal numbers of quarks and antiquarks, and antibaryons contain more antiquarks than quarks. The conservation of baryon number then becomes a consequence of the conservation of quark number because baryons will have some positive value of quark number, while mesons and leptons will have zero quark number, so that a baryon cannot change into any combination of mesons and leptons. The simplest way to make mesons and baryons out of quarks is to require that baryons consist of three quarks bound together, while mesons are a quark and an antiquark bound together, so that the baryon number is just three times the quark number. This picture is usually called the quark model of hadrons. In this model, the association of baryon number 1 with half-integer spin is a consequence of the fact that the quarks all have half-integer spin. It should be realized that because of the possibility of virtual creation of quark–antiquark pairs, the "dressed" baryon state will to some extent consist of four quarks and an antiquark, etc. However, it is believed that these deviations from the simple model do not disturb the main features of the quark model.

Some simple qualitative aspects of this model agree with the observed properties of hadrons. A bound state of two spin $\frac{1}{2}\hbar$ particles which have an orbital angular momentum of zero can have a total angular momentum of 0 or of $1\hbar$ depending on whether the spins are parallel or opposite. We should therefore expect to find mesons with spin 0 and spin $1\hbar$ among the hadrons, and it is the case that the mesons of lowest rest energy all have spin zero, while those with somewhat higher rest energy have spin $1\hbar$. There are mesons of still higher rest energy whose spin is greater than $1\hbar$ and they can be interpreted as bound states with an orbital angular momentum different from zero which can then add to the quarks' spin to produce a higher total angular momentum. The fact that the energy of the state increases as the angular momentum increases agree qualitatively with what happens in nuclei and atoms.

Similarly, when three spin $\frac{1}{2}\hbar$ particles are bound together in a state, in which all orbital angular momenta are zero, the total angular momentum can be $\frac{1}{2}\hbar$ or $\frac{3}{2}\hbar$, depending on how the spins are arranged. Again, it is known that the lowest-energy baryons all have spin of $\frac{1}{2}\hbar$, while another set with somewhat higher energy have a spin of $\frac{3}{2}\hbar$. So the angular momentum properties of the low-rest-energy hadrons can be understood fairly well in terms of the quark model, provided that the quarks are spin $\frac{1}{2}\hbar$ particles.

We have seen that some hadrons carry conserved quantities other than baryon number, i.e., hypercharge and electric charge. Since these quantities vary from baryon to baryon, and from meson to meson, their values must reflect a varying quark content of different hadrons. Because in the simplest model all baryons contain three quarks, this needed variation in quark content means that there exist different types of quarks which vary in electric charge and in hypercharge, and which are found in different combinations in each baryon and in each meson.

It is usually assumed that three types of quarks are found in the known baryons and mesons, and that in strong or electromagnetic interactions the number of each of these types of quarks, minus the number of corresponding antiquarks, remains unchanged. This conservation of the number of particles minus the number of antiparticles is similar to the known conservation laws, and is consistent with the possibility of pair creation. I shall label the three types of quark q_1, q_2, and q_3, and the corresponding antiquarks $\bar{q}_1$, $\bar{q}_2$, $\bar{q}_3$. For the hadrons constructed from these quarks there will be three conservation laws. One of these will be just the conservation of the total quark number, which we have seen is equivalent to baryon number conservation. The other two laws will forbid certain baryons and mesons from transforming into other baryons and mesons. For example, a meson made of a q_1 quark and a $\bar{q}_2$ antiquark cannot change into a meson made of a q_1 and a $\bar{q}_1$, because the number of q_1 and of q_2 would both change. Such restrictions are implied by the conservation of hypercharge and of electric charge, so the other two quark conservation laws can be identified with the conservation laws of hypercharge and electric charge that are observed among the hadrons. This identification tends to make the latter conservation laws less mysterious, in that the quantities being conserved now have a relatively simple interpretation in terms of numbers of objects of a definite kind.

In this respect, the simplification in particle physics allowed by the quark model is similar to that achieved in chemistry in the eighteenth century by the introduction of chemical elements. It was then known that definite quantities of one substance would, in many chemical reactions, always change into definite quantities of other substances. It was suggested by Antoine Lavoisier and others that these rules could be understood by assuming every substance to be composed of certain fundamental substances, the chemical elements, in specific combinations. It was further necessary to assume that the amounts of each element never changed in chemical reactions. This explanation of the chemical "conservation" laws was generally accepted after the work of Lavoisier, even though some elements, such as fluorine, were not discovered as isolated substances until much later.

The existence of three types of quark and three types of antiquark also gives a natural explanation of the observed multiplets of mesons and baryons. There are nine distinct combinations of a quark and an antiquark, which can be identified with nine mesons. These combinations are $(q_1\bar{q}_1)$, $(q_1\bar{q}_2)$, $(q_1\bar{q}_3)$, $(q_2\bar{q}_1)$, $(q_2\bar{q}_2)$, $(q_2\bar{q}_3)$, $(q_3\bar{q}_1)$, $(q_3\bar{q}_2)$, and $(q_3\bar{q}_3)$. Since the quark and antiquark can have their spins pointing parallel or opposite, there should be two such sets of nine mesons, one set of mesons each having spin $1\,\hbar$, and one set of mesons each having spin 0. A set of nine mesons of spin $1\,\hbar$, with various values of hypercharge and electric charge and with rest energies varying between 770 and 1,020 MeV, were discovered around 1960. These mesons all decay by strong interactions into lighter mesons. The similarities in their spins and decay properties make it likely that they are a family of particles related to each other. Such a relationship would be plausible if the three types of quarks have approximately equal masses and similar strong interactions, as then we would expect bound states containing the quarks to have similar properties as well. The small differences among the mesons can be attributed to corresponding small differences among the quarks.

Further evidence for this picture of the mesons comes from the fact that nine different mesons of spin 0 with the same pattern of electric charges and hypercharges as the mesons of spin $1\,\hbar$ are also known. Eight of these are the metastable mesons of Table II, while the ninth is an unstable meson of higher rest energy than those. These mesons also have similar strong interaction properties. Their rest energies are not as nearly equal to one another as those of the spin $1\,\hbar$ mesons, but are smaller than those of most other mesons and so it appears reasonable to consider the spin zero mesons as one family also. Therefore, both of the simple sets of quark–antiquark bound states actually exist in nature.

The Quark Model of Baryons

The extension of the quark model to the baryons is complicated among other reasons by the fact that three quarks are contained in each baryon. The number of distinct baryon states containing three quarks, each of which can be one of three distinct types, is difficult to calculate, and depends on the states of the quarks within the baryons. Suppose that the three quarks in a baryon were in the same quantum state, say by having zero orbital angular momentum, the same energy, and parallel spins. This would correspond, by the rules for adding angular momenta, to a baryon of spin $\frac{3}{2}\hbar$. If this situation could occur for any combination of the three types of quark, it could be done in the following ten ways: $(q_1q_1q_1)$, $(q_1q_1q_2)$, $(q_1q_1q_3)$, $(q_1q_2q_2)$, $(q_1q_3q_3)$, $(q_1q_2q_3)$, $(q_2q_2q_2)$, $(q_2q_2q_3)$, $(q_2q_3q_3)$, and $(q_3q_3q_3)$. Note that because the quarks are all in the same state, there is no difference between $(q_1q_2q_1)$ and $(q_2q_1q_1)$ as both contain the same quarks.

We therefore expect that there exists a set of ten baryons, each with spin $\frac{3}{2}\hbar$, approximately equal mass, and various electric charges and hypercharges. Ten such particles were indeed discovered by the 1960s. One of these is the metastable Ω^-, the other nine being unstable particles. These ten baryons may then be identified with the ten combinations of quarks listed above.

However, this set of ten baryons with a spin of $\frac{3}{2}\hbar$ does not include the neutron or proton which have spin $\frac{1}{2}\hbar$. These nucleons must involve three quarks whose spins are not all parallel. It can be shown that by combining three quarks it is possible to produce a set of eight such baryon states with a spin of $\frac{1}{2}\hbar$. Again, these states should have approximately equal mass, and various electric charge and hypercharge. This set of eight can be identified as the proton, the neutron, and the six other metastable baryons of Table II discovered in the 1950s. Most, although not all, of the known baryons have now been classified into families of bound states of three quarks.

However, this agreement between experiment and the quark

model raised a perplexing question. Because quarks are assumed to have spin $\frac{1}{2} \hbar$, they should obey the exclusion principle. In that case, it would be impossible for two or three quarks of any one type to be in the same state. Therefore, combinations such as $(q_1q_1q_1)$ could not occur. Physicists are not willing to abandon the exclusion principle, which has impressive theoretical arguments in its favor. Also it is hard to see how baryons could obey this principle if their quark constituents did not. Therefore, to avoid this conflict, physicists decided to complicate the quark model in the following way. Each type of quark such as q_1, is assumed to occur in three forms, sometimes called three "colors." The phrase "color" should not be thought to have anything to do with color as perceived by the eye. Instead, it is a fanciful way of describing a quality of quarks that makes no essential difference in their properties, just as the color of a pair of socks makes little difference in the function of the socks. The three "colors" of quark are assumed to be completely identical in all their observable properties, unlike the three types of quark which differ from one another in electric charge or in hypercharge. The restrictions of the exclusion principle do not apply to quarks of the same type but different color, or to quarks of different type and the same color, as these must be thought of as different objects. The exclusion principle only applies to quarks of the same type and the same color. Therefore, the combinations that previously appeared to be forbidden by the exclusion principle can now occur. For example, the combination $(q_1q_1q_1)$ is allowed, provided that the three q_1 quarks actually are of three different colors, since these will not exclude each other from the same state. It is possible to give a mathematical description of quarks with the extra property of color in such a way that those baryon states allowed by the exclusion principle are precisely the ones that are observed. In doing this, it is necessary to require that the states that occur are symmetric combinations of all colors, as otherwise there would exist more quark combinations than there are baryons. The description of mesons is not much affected by the extra complication of color, because the combination of a quark and an antiquark does not satisfy the exclusion principle anyway. However, the same requirement of a symmetric combination of colors must be made for mesons as well.

The hypothesis of color for quarks is a strange assumption, because, for the first time in subatomic physics, it introduces a distinction between particles that are identical in all observable ways, e.g., between two q_1 quarks of different color. Although present evidence seems to favor this hypothesis, it has some flavor of being *ad hoc,* and it may eventually be replaced by an alternative description.

The quark model also clarifies the reason for the success of unitary symmetry in classifying the hadrons. The basic reason for the success of this symmetry is the similarity in properties of the three types of quarks q_1, q_2, and q_3. It follows from this similarity that hadrons containing various combinations of these quarks bound together would also have similar properties, provided that the quarks in one hadron are in the same quantum states as the quarks in another hadron. Unitary symmetry is the mathematical statement of this similarity of properties. However, the quark model makes predictions of similarities between hadrons that go beyond those of unitary symmetry. For example, the existence of a ninth meson of spin $1\,\hbar$ with properties similar to the other eight is predicted by the quark model, but not by unitary symmetry, which only predicts eight such similar mesons. While the quark model was originally introduced as a mathematical device for deducing unitary symmetry, a substantial amount of evidence for its validity has been obtained, and it has taken on an independent status as by far the most convenient way to describe hadrons.

The Electric Charges of Quarks

Once the baryons and mesons have been identified with specific bound systems of quarks, it becomes possible to determine the precise electric charges of the quarks from those of the hadrons. For example, the bound system ($q_3q_3q_3$) is identified as the Ω^- baryon. Since this particle has an electric charge of $-1e$, where e is the charge of a proton, it follows that each q_3 has a charge of $-\frac{1}{3}e$, so that three together will have charge $-e$. Furthermore, the K^- meson has a charge of $-e$, and is composed of a q_3 and a $\bar{q}_1$, which with the previous information about q_3 implies that the charge of $\bar{q}_1$ is $-\frac{2}{3}e$, and that of q_1 is $+\frac{2}{3}e$. Finally, the positive pion, with charge e, is composed of q_1 and $\bar{q}_2$, so that the charge of q_2 is also $-\frac{1}{3}e$. Therefore, the quark model leads to the unexpected conclusion that the electric charges of quarks are fractions of the charges carried by all the known particles. This result is not so unreasonable, since the quarks are supposed to be different kinds of particles than the hadrons. Nevertheless, physicists were somewhat shocked when this possibility was first conjectured by Murray Gell-Mann. In spite of the fractional charges of the quarks, any combination of three quarks, or of a quark and antiquark, will have a charge that is an integer times e. For example, a proton consisting of $(q_1q_1q_2)$, will have a charge of $(+\frac{2}{3}e)+(+\frac{2}{3}e)+(-\frac{1}{3}e)$, which is e. A neutron consists of $(q_1q_2q_2)$, and these quarks have charges adding to zero. As a result, none of the quarks can transform completely into ordinary hadrons, both because of the conservation of electric charge, and because of the conservation of quark number. This does not imply that all types of quark are stable. Only the quark with the lowest rest energy will be stable. The other quarks, which may have slightly higher rest energy in order to account for the differences in rest energy among similar baryons, could decay by weak interactions into the lowest-energy quark, together with mesons or leptons. The precise pattern of decays would depend on which quark has the lowest rest energy, and on the differences in the energies. Note that the difference in electric charge between any two types of quark are integer multiples of e, so that one kind can convert into another by the emission of ordinary particles.

If unbound stable quarks with fractional charge occur naturally, or can be produced, they should be easy to detect. The response of various detecting instruments to a subatomic particle passing through them depends on the electric charge of the particle. For instance, the amount of ionization, and hence the thickness, of a track produced in a bubble chamber increases as the charge increases. Thus if pairs of free quarks are produced in collisions among hadrons, there should sometimes be an especially thin track in a bubble chamber in which high-energy hadrons have been allowed to collide with the atoms in the chamber, corresponding to the fractional charge of a quark passing through the chamber. Various searches for events of this type have been carried out, but no indications of quarks have been seen thus far.

When the quark model was first proposed, it was considered plausible that the quarks themselves would at some point be observed in an unbound state, either through their natural occurrence or through their production in collisions between hadrons. Since particles of fractional charge and with mass comparable to that of the known hadrons had not been found in either of these situations, it was imagined that unbound quarks were very rare in nature, and also that their rest energy was sufficiently large, say greater than 5 GeV, that they would not be produced in collisions at the energies then available. This explanation has come to appear rather implausible to most physicists. Furthermore, some of the most interesting successes of the quark model in explaining properties of hadrons seem to require that the quarks bound inside hadrons have a rather low mass, perhaps ⅓ that of the nucleon. Therefore, physicists have looked for alternative explanations of the fact that unbound quarks have neither been discovered nor produced.

How Quarks Are Bound Together

The question of whether quarks can be observed as free particles must be considered together with the question of how quarks are bound together to form hadrons. The quarks cannot be bound by the same force that binds nucleons into nuclei, since the force that binds nucleons is mostly due to the exchange of pions between the nucleons, and the pions themselves exist because some force acts between the quarks. There must be some other kind of strong interaction acting between the quarks. One possibility is that some other, yet unobserved particles are virtually exchanged between quarks to produce the force needed to bind quarks together into hadrons. Such hypothetical particles have been called "gluons," and their properties have been studied by physicists. In this model the new fundamental strong interaction is that between the quarks and the gluons, and all other strong interactions are indirect manifestations of that interaction, in somewhat the same sense as all chemical interactions are a manifestation of the fundamental electromagnetic interaction between charges and photons. While this approach may prove to be correct, it begins again to bring into physics an unfortunate multiplicity of types of particles, and there is little evidence as of 1976 for the existence of "gluons." Other mechanisms for binding quarks have been proposed, but none of them has proven completely convincing, and the problem remains an open one.

Leaving aside the question of the ultimate origin of the force between quarks, we can inquire about the properties of that force. One possibility is that a force acts between quarks which—unlike all other forces that have been discovered in physics until now—does not decrease as the distance between the quarks increases. If such a force exists, it would be impossible for two quarks to separate indefinitely from one another, because the force between them would always eventually overcome any impulse that had driven the two apart and would reunite them. Therefore, quarks would never be found singly, at least at macroscopic separations from other quarks, and the problem of why they have not been observed as isolated particles would be avoided. However, a problem exists with the suggestion of an nondecreasing force. It is known that no such force occurs between ordinary hadrons, since hadrons can exist in isolation. However, the various quarks in one such hadron presumably exert forces on the quarks in another hadron, as well as upon each other. In order to avoid forces that fail to decrease with distance between two hadrons, the forces between all the quarks in one hadron and all the quarks in the other hadron must somehow cancel each other. Such a cancellation might be the result of a mechanism similar to that which occurs for electrical forces. In that case two charges exert a relatively large force on each other, while two neutral bodies, composed of equal amounts of positive and negative charge, exert much smaller forces on each other, because the forces between the different types of charge in each body tend to cancel. Some models of the interactions between quarks have been proposed in which such a cancellation among attractive and repulsive forces between the quarks in different hadrons occurs. These models also help to explain why bound systems of two or four quarks are not observed, in spite of the strong binding forces between quarks. The reason for this is that a two- or four-quark system does not contain the right combination of quarks to arrange a cancellation of the forces between two such systems that do not decrease with distance. Therefore, any two- or four-quark systems that existed would be pulled together by these forces, and also could not be observed as separated objects. Only sets of 3, 6, 9, etc., quarks, or equal numbers of quarks and antiquarks, can have the correct combination of attractions and repulsions to give a decreasing force, and hence only such combinations can appear as separated objects. These sets are just the ones that have been identified as baryons and mesons. In the version of the quark model in which each type of quark occurs in three colors, the quark combinations that can occur as separated objects are just those in which the different colors appear symmetrically, and this works out to be the sets mentioned.

If this explanation is correct, the quarks in a hadron may be relatively light particles, and move rather slowly inside the hadron. The picture we would then have is that a hadron would be like a box with rigid walls, which confines a small number of particles (Fig. 20). When the particles are away from the walls, the forces that act on them are small, and the particles behave as if they were free. Only when the particles approach the wall does a strong force act on them, the main effect of which is to confine them to the interior of the box. More sophisticated versions of this idea have been worked out and described as the "bag" model of hadrons. Two problems arise from this description: when two hadrons collide, their "bags" must for a short time overlap and then reform as individual containers, and it is not clear how this happens. Also, it is not known how this description can be made consistent with the requirements of relativistic quantum theory. On the other hand, detailed calculations have shown that several aspects of the behavior of hadrons can be understood in terms of this picture. The agreement between some experiments and calculations done on the basis of this form of the quark model may be taken as some confirmation of the accuracy of the model, and indirectly, as evidence for the existence of quarks within hadrons.

This approach to the quark model is presently one of the major areas of research in subatomic particle physics.

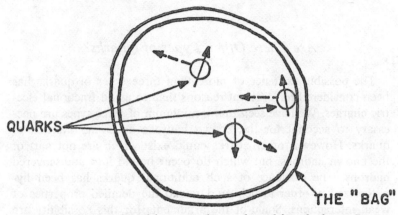

FIGURE 20. One model of how quarks are bound into hadrons. In one version of the quark model of hadrons, fancifully called the "bag" model, the quarks in a hadron move about freely inside a container, but are strongly repelled by the walls of the container, and so do not emerge. A baryon "bag" containing three quarks, indicated by circles with arrows, is shown in the figure. The nature of the bag itself is unclear at present.

It is also possible that quarks have not been observed for other reasons—including that the quark model is incorrect. The possibility exists that the electric charge of quarks is integral, rather than fractional. One way in which this could be arranged is that there are more than the three different types of quarks we have discussed present in hadrons. For example, if there were six types, it would be possible for their charges to be integral and nevertheless have all the charges of the known baryons take on the observed values. In this case, the quarks would not have been detected in experimental searches for fractionally charged particles, but would instead appear as new stable particles of integral charge. If such quarks existed as free particles with a rest energy of less than 1 GeV, they would probably be readily produced in pairs in hadron collisions, and detected in standard bubble-chamber searches for new particles. Hence even integrally charged quarks, if they exist unbound, would have to have rather high masses. This possibility has not been pursued very much.

Are There Other Types of Quarks?

The possible existence of more than three types of quarks has been considered for different reasons than to avoid fractional electric charges. We have seen that a minimum of three types are necessary to account for the known hadrons as bound systems of quarks. However, other quarks could exist which are not part of the known hadrons, but which do occur bound into undiscovered hadrons. The existence of such additional quarks has been hypothesized in order to help understand the detailed properties of weak interactions. Some of the arguments for this possibility are discussed in the next chapter. In one such model, originally suggested by James Bjorken and Sheldon Glashow, there exists a fourth type of quark, with a charge of $+\frac{2}{3}e$, in addition to the three fractionally charged quarks already mentioned. It is assumed that the number of the new quarks, minus the number of antiquarks, is conserved in strong and electromagnetic interactions, just as is the case for the other types of quarks. This is sometimes described by saying that the new quarks carry a conserved property called "charm." Any hadron containing a different number of "charmed" quarks and "charmed" antiquarks would also carry a nonzero value of this property of "charm." It is imagined that hadrons could be made of combinations of "charmed" quarks or antiquarks with ordinary quarks or antiquarks. It is also assumed that the "charmed" quark has a substantially higher rest mass than the other three quarks, which would tend to make hadrons containing one or more "charmed' quarks higher in rest mass than hadrons containing only the other quarks. Both baryons and mesons containing "charmed" quarks are expected to exist. These "charmed" hadrons could be produced in pairs in collisions between ordinary particles, provided that the total "charm" of the particles produced adds up to zero. The least massive "charmed" hadrons would decay, via weak interactions, into ordinary hadrons, as it is expected that "charm," like hypercharge, is not conserved in weak interactions. The half-lives of these "charmed" hadrons would be something around 10^{-13} sec. This is too short for them to leave visible tracks in ordinary detectors, but it should be possible to identify them through their distinctive decays. Evidence for the existence of "charmed" mesons, decaying into K mesons and pions, was found in 1976. Also, evidence for "charmed" baryons, decaying into lambda-zero baryons and pions, was found. These discoveries make the hypothesis of "charmed" quarks seem very probably true. It is also expected that more massive "excited states" of "charmed" hadrons exist, which would decay by strong interactions into the less massive "charmed" hadrons.

Somewhat earlier, in late 1974, indirect evidence for the hypothesis of "charmed" quarks was discovered in the form of several mesons which could be bound systems of a "charmed" quark and a "charmed" antiquark. In such a bound system, the total value of the "charm" is zero, as it is equal and opposite for quark and antiquark. Similarly, the electric charge of such a meson is zero. The mesons that were discovered are indeed electrically neutral, and from their observed properties cannot carry "charm," because they are produced singly by electromagnetic interactions. Their rest energies lie between 3.1 and 3.7 GeV. This is consistent with the expectation that "charmed" quarks and bound systems containing them are heavier than other particles. A surprising property of these mesons is that they do not easily transform into ordinary hadrons. Such transformations, for example in the form of a decay of the heavier mesons into lighter mesons such as pions, occur only 10^{-3} times as often as would be expected for ordinary hadrons of such high rest energy. This is puzzling because as the total "charm" of the heavy mesons is zero, they are not inhibited by any known conservation law from decaying into lighter mesons. It remains to be seen whether the large suppression of such decays can be understood within the model in which the heavy mesons are bound systems of "charmed" quarks, or whether some further hypotheses are needed. However, many of the details of the energies, spins, and other properties of these mesons do emerge in a natural way from this model, and most physicists think that some version of it is correct.

The apparent success of the prediction of "charmed" quarks

has led many physicists to hypothesize still more types of quark with still higher rest energies. It remains to be seen whether any of these other suggestions will be equally successful.

In summary, the quark model has been very successful in describing the quantum numbers and other properties of individual hadrons. It has been less successful in describing strong interactions between several hadrons, which is admittedly a more difficult problem. The proliferation of quarks that has been necessary to describe the phenomena has perhaps made the model itself appear to be a less likely candidate for an ultimate theory of the subatomic particles, but no phenomena that would take us beyond quarks to some deeper level have yet been identified. It is conceivable that an alternative to the quark model for explaining the hadrons will eventually turn out to be correct. Some such proposals have been made, but have been a good deal less successful than the quark model in explaining what is known.

If the quark model is correct, then the known hadrons themselves have a complex inner structure, much as has been found previously to be the case for atoms and for nuclei. For that reason, the name particle has perhaps become somewhat inappropriate for the hadrons, and should probably be reserved for the quarks, leptons, and any other objects not made of quarks. The relation between the quarks and the leptons remains to be clarified.

X

Weak Interactions and Symmetry Principles

The interaction responsible for the decay of the pion and the decay of the muon is quite different in strength than the strong interactions. This cannot simply be inferred from the relatively long lifetimes of these particles, because those lifetimes depend also on the release of kinetic energy and on the number of particles which emerge from the decay. The laws of quantum mechanics enable us to correct for those factors, and infer a kind of relative strength of various interactions. When this is done, it is found that several interactions—first discovered in the decays of various particles—have roughly comparable strengths to one another, all much smaller than those of the strong hadron interactions discussed in Chap. VIII. These include the interaction responsible for charged pion decay, the interaction involved in muon decay, the interaction involved in neutron beta decay, and the interaction involved in the decays of "strange" hadrons to ordinary hadrons (Table III). This collection of interactions possessing a roughly common and small strength is called the weak interactions.

The weak interactions are now believed to act among all of the subatomic particles, leptons, mesons, and baryons. However, most of the hadrons decay into other hadrons via strong interactions so rapidly that there is insufficient time for the weak interactions to act in an observable way on them. Only for the leptons and the lightest of the hadrons, among which decays by strong interactions are impossible, can the weak interactions be observed in decays. The weak interactions have the interesting property of acting with approximately the same strength between many types of particles. In this respect they are like electromagnetic interactions which are the same for all particles of the same charge. This similarity is made use of in the "unified" theories of weak and electromagnetic interactions that we shall discuss.

TABLE III

Some Weak Decay Processes

DECAY PROCESS	HALF-LIFE (SECONDS)	DECAY STRENGTH
negative muon → electron + mu-neutrino + e-antineutrino	1.5×10^{-6}	10^{-10}
negative pion → negative muon + mu-antineutrino	1.8×10^{-8}	1.5×10^{-12}
Neutron → proton + electron + e-antineutrino	657	10^{-10}
Lambda zero → proton + negative pion	1.8×10^{-10}	2×10^{-12}

The half-lives for these decay processes vary considerably, although they are all much larger than the 10^{-22} sec or less associated with strong decays. However, when allowance is made for the big difference in kinetic energy released, and for the number of particles in the decay, the corrected decay strengths are much more similar, as shown in column three.

Although originally observed in decays, the weak interactions can also be studied in scattering processes, especially those of neutrinos. It is possible to observe the "inverse" of the beta decay by scattering a beam of neutrinos from neutrons, producing protons and electrons. By varying the energy of the incident neutrinos, one can study weak interactions over a wider range of energies and among other particles than are available in decays.

Although the different weak interactions were originally identified on the basis of a common strength, it was soon found that they had other properties in common. In 1956–1957, Tsung-Dao Lee and Chen-Ning Yang proposed that certain symmetry principles previously thought to be universally valid were in fact not satisfied by the weak interactions as a class. That suggestion was soon demonstrated experimentally by Chien-Shung Wu, Leon Lederman, and others. Furthermore, it has been possible to account for the differences in strength that do exist within the class of weak interactions in terms of some very simple assumptions about the fundamental nature of the weak interactions. Finally, in recent years theories have been proposed that relate the weak interactions to the more familiar electromagnetic interactions of charged particles with photons. In these "unified" theories, some light is shed on why the strength of the weak interactions is so small compared to strong or electromagnetic interactions. On the whole, the theory of weak interactions has seen the most progress of any area of theoretical study of subatomic particles in the years 1950–1975.

The Scattering of Neutrinos

In order to understand the characteristic properties of weak interactions, let us compare a process involving weak interactions to another involving strong interactions. For the former we take neutrino–proton scattering, for the latter, the proton–proton scattering discussed earlier. Both processes can be studied at energies of tens of GeV, where the protons are traveling at almost the velocity of light, just as the neutrinos do. In each case, the typical reaction produces many hadrons.

In the neutrino reaction, a lepton always remains, either a neutrino or a charged lepton. The type of charged lepton depends on the way the incident neutrino was originally produced. The incoming neutrinos may have been produced with either positive or negative muons (in the decay of pions) or with positive or negative electrons (in certain decays of K mesons). The neutrinos produced with muons always make muons, and those produced with electrons always make electrons. This fact, discovered in 1960, indicates that there are two distinct types of neutrino, called muneutrinos and e-neutrinos, each distinguished by some property that also distinguishes muon from electron. This property cannot be the mass in the case of the neutrinos, as both rest masses are zero within experimental accuracy. Instead, from a lack of anything better, physicists ascribe to muons and mu-neutrinos a conserved additive property called muon-number, and to electrons and e-neutrinos a similar property called electron number. The values of these numbers for the antileptons are the negative of the values for the leptons. These properties are conserved just as is the baryon number, that is in all interactions. Another relation that is found in the experiments is that the neutrinos produced with positive leptons always make negative leptons, while those produced with negative leptons always make positive ones. Again, this indicates a difference between two types. This difference is the standard distinction between particle and antiparticle. The neutral particle produced with a positive lepton is a neutrino; that produced with the negative lepton an antineutrino. A neutrino can then only convert into a lepton particle, rather than an antiparticle, and so produces the negative lepton, which is the particle. The distinction between neutrino and antineutrino is not through electric charge, as this vanishes for both. Instead, the two are different in the relation between their spin and momentum, as we shall soon see.

The main quantitative difference between this neutrino–proton scattering and proton–proton scattering is that the cross section for the former is smaller by 10^{-11} at comparable energies. Here the term weak interaction is literally applicable. A neutrino of 10 GeV must on average pass through 10^{13} cm of matter before it has a 50% chance of giving an interaction, whereas a proton of the same energy interacts once in passing through 10^2 cm of matter. The comparison is even more marked at lower energies, because the neutrino scattering cross section decreases more rapidly as the energy decreases. It is estimated that a neutrino of 1 MeV energy could travel through one light year of ordinary matter without interacting once. In experiments done to detect the scattering of neutrinos, many tons of matter are used in the detector, but each neutrino still has less than one chance in a million of interacting. Only by using very intense beams of neutrinos can any interactions be observed. For a long time this made it impossible to detect neutrinos at all, and their existence was a hypothesis made to enforce the various conservation laws. The hypothesis was made real by showing that neutrinos could be absorbed to produce charged leptons.

Neutrino–proton scattering differs from proton–proton scattering in other ways than in the size of the cross section. The neutrino scattering cross section increases proportionately to the energy of the neutrino, at least up to the highest energies that have been measured (several hundred GeV). On the other hand, proton–proton scattering gives an approximately constant cross section from 1 GeV upward, and other hadron–hadron scatterings behave similarly. The interpretation of this difference is that the weak interactions are characterized by a range that is much smaller than that of the strong interactions which we have learned is about the range of nuclear forces, or 10^{-13} cm. The neutrino scattering cross section is not expected to stop increasing until an energy is reached that corresponds—by Heisenberg's relation that distance is inversely proportional to momentum—to the range of the weak interactions. At present, this is believed to be a distance of about 10^{-15} cm corresponding to neutrino scattering from a stationary proton at an energy of 700 GeV or so. Experiments at such high energies are not yet practical, and so the range of the weak interactions, if any, must be established by other methods.

Yet another way in which neutrino–proton scattering differs from proton–proton scattering is with respect to the conservation of certain quantities characterizing the hadrons. Both of these processes conserve energy, momentum, electric charge, baryon number, and the muon and electron numbers. However, we have seen that in strong interactions, some additional symmetries are present which imply the conservation of other quantities such as hypercharge. This implies, for example, that in proton–proton scattering it is impossible to produce a single K meson along with pions and nucleons. In neutrino–proton scattering, on the contrary, hypercharge is not conserved, and single K mesons can be, and are, produced with a cross section that is a few per cent of the total scattering cross section. In other words, hypercharge is almost but not quite conserved in weak interactions. A similar inference can be drawn from a study of weak interaction decays. It is not known at present why this is the case.

Symmetry and Asymmetry in Weak Interactions

Two much more spectacular examples of ways in which weak interactions differ from strong interactions were discovered in 1956 after a proposal by Lee and Yang. I will first describe how these were detected in weak decays. A positive pion at rest usually decays into a positive muon and a mu-neutrino traveling in opposite directions. Each of these particles has spin $\frac{1}{2}\hbar$, which could point either parallel to or opposite the direction of motion. It is possible to measure the spin direction of the muon. This is done by allowing the muons to scatter from electrons whose spins have been fixed by the action of an external magnetic force. The force between muon and electron, and hence the scattering cross section, depends on whether the muon and electron spins are parallel or opposite, so a measurement of this cross section determines the direction of the muon spin. When this was done, it was found that the positive muon always has its spin pointing opposite to its direction of motion. Since the pion has zero spin, the mu-neutrino must by the conservation of angular momentum also have its spin pointing opposite to its momentum (Fig. 21). Particles whose spin points opposite to their momentum are called left handed (because their motion resembles that of a left-handed screw). Those whose spin points in the direction of motion are called right-handed. In the corresponding decays of negative pions, the negative muons and the antineutrinos are found to be right handed. Experiments on neutrinos originating in various reactions such as beta decay, muon decay, etc., confirm that the neutrinos emitted are always left handed, antineutrinos always right handed. The charged antileptons emitted with the neutrino are, however, usually right handed like the antineutrino; the charged leptons usually left handed, similar to the neutrinos. The pion decay into a muon and the other two particle decays are anomalous in the respect that the conservation of angular momentum forces the antilepton to have the same handedness as the neutrino. In other decays, the conservation laws do not force this to occur, and the charged antileptons

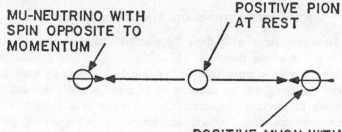

MU-NEUTRINO WITH
SPIN OPPOSITE TO
MOMENTUM

POSITIVE PION
AT REST

POSITIVE MUON WITH
SPIN OPPOSITE TO
MOMENTUM

FIGURE 21. Decay of a pion by weak interactions. A positive
pion at rest, indicated by an open circle, decays into a positive
muon, and a mu-neutrino, each indicated by a circle with an
arrow. The muon and the neutrino each have a definite energy and a spin which points in the direction opposite to the
momentum.

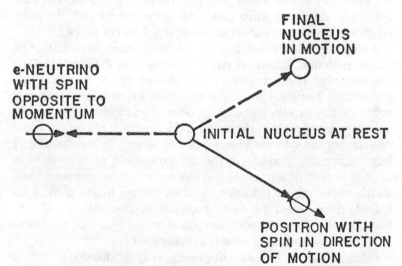

FINAL
NUCLEUS
IN MOTION

e-NEUTRINO
WITH SPIN
OPPOSITE TO
MOMENTUM

INITIAL NUCLEUS AT REST

POSITRON WITH
SPIN IN DIRECTION
OF MOTION

FIGURE 22. Positron beta decay of a nucleus by weak interactions. A
proton in a nucleus, indicated by an open circle, converts into a neutron, so the nucleus changes to another nucleus, indicated by an open
circle in motion. A positron and an *e*-neutrino are emitted. Because
three particles occur in the decay products, the energy of each is
indefinite, and only the sum of the three energies is fixed. The neutrino
spin points opposite to the direction of motion, while the positron spin
tends to point in the direction of its motion.

are free to emerge with their "preferred" handedness, which is opposite that of the neutrinos (Fig. 22).

These simple observations imply that two symmetry principles previously thought to be universally true are actually not satisfied by weak interaction processes. One of these is a symmetry between particles and antiparticles known as charge-conjugation symmetry, symbolized as C. The symmetry between particle and antiparticle that is not satisfied in weak interactions is a simple substitution of antiparticles for particles everywhere in a reaction; no other changes in the properties of the particles, such as momentum and spins occur (Fig. 23). This clearly fails in pion decay, since the positive pion emits neutrinos with spin opposite to their momentum, while the negative pion, its antiparticle, emits antineutrinos with spin parallel to their momentum. In order for the symmetry to be satisfied, the positive pion would have to emit neutrinos and the negative pion emit antineutrinos with the same relation between spin and momentum; this does not happen.

Another symmetry that fails in the weak interactions is that of mirror reflection, known as *parity symmetry,* or P (Fig. 24). The breakdown of parity symmetry is indicated by the very fact that the emitted neutrinos have a preferential handedness. Imagine a neutrino with its spin pointing opposite to its motion. As seen in a mirror perpendicular to the motion, the direction of motion would reverse, but the spin would not—as can be seen by watching one's finger move in a mirror—so the neutrino would seem to be right handed instead of left handed. If the weak interactions were symmetric under mirror reflection, without change in the particle involved, there would be equal probabilities for emission of left-handed and of right-handed neutrinos in the decay of a positive pion, again contradicting what is actually seen.

Indeed, as far as we know at present, only left-handed neutrinos and right-handed antineutrinos occur in nature, and can be produced, absorbed, or scattered. This property of neutrinos is connected with the fact that they have no rest mass. If the neutrino had a nonzero rest mass, it would be possible to change its handedness by moving sufficiently quickly to overtake the neutrino. This would have the effect of changing the sense of the neutrino momentum, but not of the spin, so that the neutrino would change

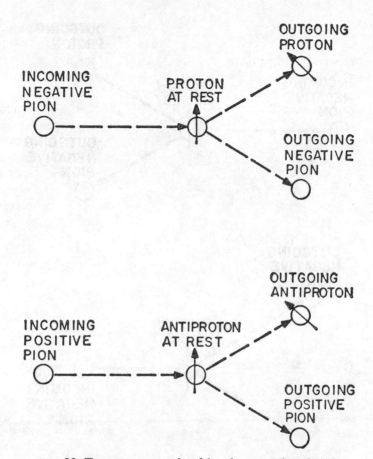

FIGURE 23. Two processes related by charge-conjugation symmetry. The diagram indicates the process of elastic scattering of a negative pion by a proton, and the process related to the first by charge conjugation—the scattering of a positive pion by an antiproton. The momentum of the particles, indicated by dotted arrows, and the spins, indicated by arrows through the circle representing each particle, are unaffected by charge conjugation, but each particle is replaced by its corresponding antiparticle in the charge-conjugate process. The processes shown involve strong interactions, which satisfy charge-conjugation symmetry, so the cross sections of the two processes are equal.

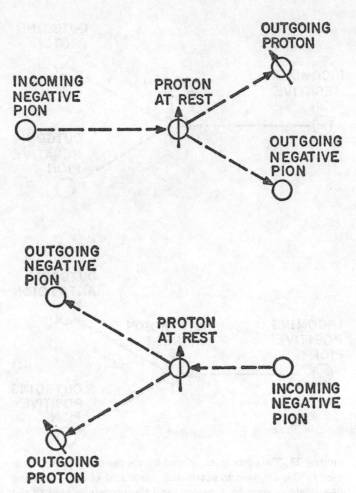

FIGURE 24. Two processes related by parity symmetry. The diagrams show the scattering of a negative pion by a proton, and the process obtained by parity symmetry. The particles involved are the same in the two processes, and their spin directions are unchanged, while the arrow representing each momentum direction is reversed. The two processes both occur by strong interactions, which satisfy parity symmetry, so the cross sections must be equal.

from left handed to right handed. However, this is impossible for massless neutrinos, which move at the speed of light and so cannot be overtaken. However, a unique handedness is not automatically associated with massless particles, as both left-handed and right-handed photons do occur in nature, corresponding to the two possible circular polarizations of the light. The charged leptons, which have a nonzero rest mass, occur with either handedness, but the weak interactions tend to involve only one handedness for each lepton, except where the other handedness is required by the conservation laws, as in pion decay.

All of this is in contrast to a typical strong or electromagnetic interaction process, in which both C and P symmetry are satisfied. For example, if an electron pair is produced by a photon, on average there will be equal numbers of negative electrons that are left handed or right handed and equal numbers of positions of each handedness. Thus the particles and antiparticles behave similarly, and neither has a preferred handedness in conformity with the symmetry requirements. Experiments on a variety of strong and electromagnetic processes have found that these processes satisfy C and P symmetries to a high degree of accuracy.

It might be thought that the failure of these symmetries in weak interactions is a peculiarity of neutrinos or leptons. However, there are weak interaction processes not involving leptons in which C and P symmetries fail as well. For example, the Λ^0 particle decays by weak interactions into a proton and negative pion. The proton emitted is found to be mostly left handed, although not as completely so as neutrinos are. This implies that P symmetry is not satisfied in the decay. More indirect evidence also shows that C symmetry is not satisfied in such nonleptonic weak decays. The breakdown of the symmetry is therefore something intrinsic to the weak interactions, rather than a property of leptons alone.

For a while, it was thought that the weak interactions did satisfy a symmetry involving the combined operation of C and P, called CP symmetry. According to this symmetry, physical processes would happen in the same way if particles were replaced by antiparticles, and simultaneously, left-handed objects interchanged with right-handed ones (Fig. 25). CP symmetry is implied by the separate validity of C and P symmetry, but not conversely. The weak decays we have described are consistent with *CP* symmetry. For example, this symmetry relates the decay of a positive pion to that of a negative pion, and requires that the handedness of the neutrino and positive muon created in the former decay be oppo-

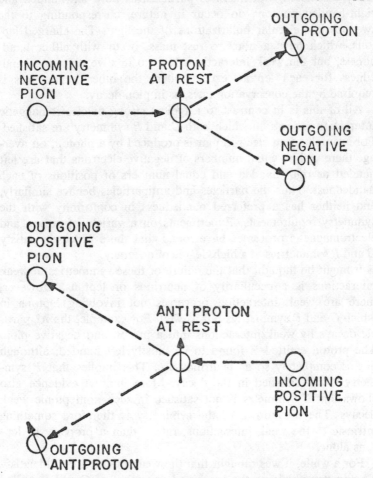

FIGURE 25. Two processes related by *CP*, the combination of parity, and charge conjugation symmetries. Each particle in the first reaction is replaced by its antiparticle in the second reaction, and also each momentum direction is reversed. Because the reactions satisfy *C* and *P* symmetries separately, they also satisfy *CP* symmetry, as can be seen from Figs. 23–25.

site to the handedness of the antineutrino and negative muon created in the latter decay, in agreement with observations. The properties of the neutrino with respect to handedness are also consistent with *CP* symmetry, and this symmetry is satisfied to a high degree of accuracy in almost all known processes: strong, electromagnetic, and weak. One exception is now known, and will be discussed below.

Still another symmetry relevant to weak interactions is *TCP* symmetry. This is a combination of *CP* with a symmetry known as time-reversal symmetry, or *T*. Time reversal does not imply a literal reversing of the direction of time. Instead, it is a comparison between two physical processes that are related by having all velocities and spins reversed in direction and the objects in the initial state interchanged with those in the final state (Fig. 26). For example, in the decay of a positive pion into a muon and neutron, the time-reversed process would be the combination of a muon and neutrino to produce the pion. The rule of changing momenta and spins is such that the handedness of the particles is unchanged between the two processes being compared (see Fig. 26). Hence the time-reversed process of pion decay is a possible physical process, and the symmetry implies that it would occur in the same way as the decay if the particles would be brought together.

The processes compared by time-reversal symmetry might be thought of as the two ways that any process would appear as viewed in an ordinary motion picture of the process and as viewed in a motion picture that is run backward. For most large scale phenomena, it would be easy to distinguish the two pictures because of the irreversible nature of most macroscopic physical effects. However, this kind of irreversibility is not usually thought to characterize atomic and subatomic processes, and time-reversal symmetry is an expression of that fact. The same weak interactions that satisfy *CP* symmetry are also found to satisfy time-reversal symmetry, although arguments leading to this conclusion are somewhat indirect and mathematical. There is also substantial evidence that strong and electromagnetic interactions satisfy time-reversal symmetry, but this symmetry fails to be satisfied in the same circumstance as does *CP* symmetry.

It was no surprise to physicists that *CP* and *T* symmetries are correlated in their validity or invalidity. This could be inferred from a result known as *TCP* symmetry believed to be universally

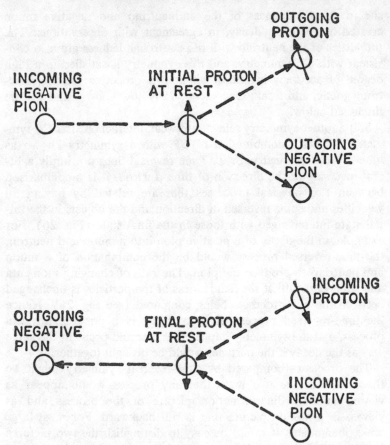

FIGURE 26. Two processes related by time-reversal symmetry. The diagram shows the scattering of a negative pion by a proton and the process obtained by time-reversal symmetry. Those particles that are incoming in the first process are outgoing in the second process, and conversely. Also, all momenta and all spins have reversed directions. The processes shown involve strong interactions, which satisfy time-reversal invariance, so the cross sections of the two processes are equal.

true. As the symbol suggests, this principle states a relation between two physical processes for which all three of the operations T, C, and P are applied to one process to obtain the other. Thus in applying the TCP operation to positive pion decay, we obtain the process of a negative muon and antineutrino with their proper handedness colliding to make a negative pion. It has been found through mathematical analysis that in relativistic quantum theory *TCP* symmetry should hold for all physical interactions, whether or not the three individual symmetries are satisfied. This implies that if any one of the three is not satisfied, then at least one of the other two must also fail, as is the case for *C* and *P* symmetries in the weak interaction. Finally, it implies that if *CP* invariance fails, then *T* symmetry must also fail, or else the product *TCP* could not be satisfied (Fig. 27).

TCP symmetry also has some consequences of its own. In particular, it implies that the mass of any particle and its antiparticle must be equal. One way to see this is to recognize that the mass of a particle is a quantity that can be assigned to a single particle at rest. The *TCP* operation applied to a particle at rest yields an antiparticle at rest. If these two situations are to appear similar, the masses must be equal for the particle and antiparticle. Accurate measurements of the masses of particle and antiparticle have confirmed this prediction. *TCP* symmetry also implies that the particle and antiparticle have opposite values of electric charge, baryon number, hypercharge, etc., and that if they are unstable particles, they have equal lifetimes. Measurements of lifetimes of various unstable particles and antiparticles have also confirmed this result. *TCP* symmetry is believed to be the most fundamental of the symmetries involving discontinuous changes in the properties of particles, and has the same basis in relativistic quantum theory as the connection between spin and statistics. It is therefore reassuring that available evidence favors the validity of this symmetry in all cases.

The one known case in which *CP* symmetry is not satisfied is the decay of the particles known as neutral *K* mesons. Because the *K* mesons carry hypercharge, the electrically neutral particles called K^0 are not identical to the antiparticles called $\overline{K}^0$, as their hypercharges are different. This implies that in certain strong interaction reactions, a single K^0 can be produced, together with other particles, but a single $\overline{K}^0$ cannot be produced with the same particles, because hypercharge would not then be conserved. For example, in a negative pion–proton collision, a Λ^0 and a K^0 can be produced, but not a Λ^0 and $\overline{K}^0$. However, in the subsequent decay of the K^0 meson, which occurs by a weak interaction,

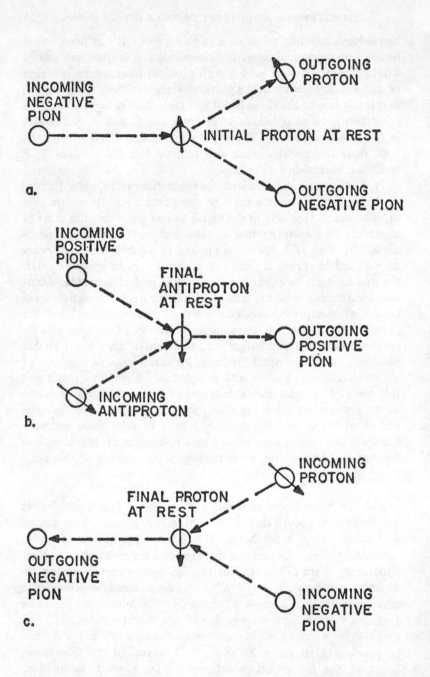

INCOMING
NEGATIVE
PION

OUTGOING
PROTON

INITIAL PROTON AT REST

OUTGOING
NEGATIVE
PION

a.

INCOMING
POSITIVE
PION

FINAL
ANTIPROTON
AT REST

OUTGOING
POSITIVE
PION

INCOMING
ANTIPROTON

b.

INCOMING
PROTON

FINAL PROTON
AT REST

OUTGOING
NEGATIVE
PION

INCOMING
NEGATIVE
PION

c.

hypercharge is not conserved. Both the K^0 and $\overline{K}^0$ decay into states of hypercharge zero such as two or three pions, or states involving both hadrons and leptons for which hypercharge is not well defined. In fact, it is possible for K^0 and $\overline{K}^0$ to make transitions, real or virtual, into the same states. This possibility, together with the fact that K^0 and $\overline{K}^0$, being particle and antiparticle, have equal mass, leads to an interesting behavior in their decay. A K^0 or a $\overline{K}^0$ do not have a unique half-life in their decay as do other unstable particles. Instead each of these decays as an approximately equal mixture of two states with two distinct half-lives. The two states with definite half-lives are called K^0 short and K^0 long, or $K^0{}_S$ and $K^0{}_L$, and the half-lives are about 10^{-10} and 10^{-7} sec, respectively. Also, the rest energies of $K^0{}_S$ and $K^0{}_L$ differ slightly by about 3×10^{-6} eV, an almost infinitesimal amount compared to the energies themselves which are about 5×10^8 eV.

If CP symmetry were satisfied in K meson decays, the $K^0{}_S$ and $K^0{}_L$ would have completely distinct decay modes. For example, the $K^0{}_S$ would decay predominantly into two pions, in a state which is unchanged by the CP operation (Fig. 28). The $K^0{}_L$ could not decay into this state, but would instead decay into a state with three pions. This would incidentally account for the great disparity in lifetimes, as it is usually true that lifetimes increase rapidly as more particles are involved in the final state.

However, it was discovered in 1964 that the $K^0{}_L$ actually does sometimes decay into two pions at a rate about 4×10^{-6} that of the $K^0{}_S$. This was indirect evidence for a small deviation from CP symmetry in the decay. A somewhat more direct indication was discovered several years later when it was found that the $K^0{}_L$ decays at slightly different rates into two modes that are related by CP, i.e., slightly more into a state containing positive pion, electron, and antineutrino than into a state containing negative pion,

FIGURE 27. Three pion–proton scattering processes. According to TCP symmetry the processes shown in a. and b. must have the same cross section. According to CP symmetry, the processes shown in b. and c. have the same cross section. Therefore, if both of these symmetries are true, the processes shown in a. and c. must have the same cross section. By comparison with Fig. 26, it can be seen that this is also the implication of time-reversal symmetry. This is an example of the result that if TCP symmetry is true, and CP symmetry is true, then T symmetry must be true.

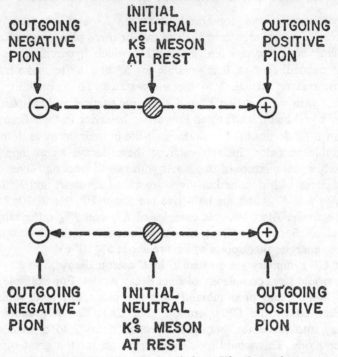

FIGURE 28. *CP* symmetry applied to $K^0{}_S$ decay into two pions. Diagram a. shows the decay of one of the neutral K meson states with a definite lifetime $K^0{}_S$, into a positive and negative pion. When the symmetry *CP* is applied to this decay, the positive pion becomes a negative pion, the negative pion a positive pion, and the direction of each momentum is reversed. The result of this is that there is no change in the final pion state at all. If *CP* symmetry were valid for this decay, it would imply that the original $K^0{}_S$ was also unaffected by this symmetry.

positron, and neutrino. If *CP* were exactly valid, then $K^0{}_L$ or $K^0{}_S$ would have to decay equally into any two modes related by *CP*. Note that this is not required by *TCP* symmetry, which relates a decay rate to a rate at which the antiparticles come together in the inverse process.

The fact that more electrons than positrons occur in $K^0{}_L$ decay has an interesting consequence concerning the meaning of particle and antiparticle. We have seen that the surplus of electrons and protons over positrons and antiprotons, in the matter on Earth and in our galaxy has no known fundamental cause. It is conceivable that in other parts of the Universe the situation is reversed, and there are more positrons, antiprotons, etc., in the matter making up the region. The question arises as to how we could communicate, say via radio signals, to a being inhabiting another region that we have a surplus of electrons rather than positrons. Since all of the gross properties of matter and antimatter (as defined by us) are symmetric, a distinction cannot be made on that basis. Of course, if a small sample of our matter could be transported to the other beings, it would be easy for the beings to determine whether their matter was made of what we call particles, or of antiparticles, by seeing whether our matter annihilates theirs. However, if we assume that only communication via photons is possible, that procedure cannot be used, as photons are identical to their antiparticles and cannot distinguish between the two possibilities.

If CP symmetry were exactly satisfied, this distinction could not be communicated by using weak interactions either. Suppose, for example, that we tried to communicate the difference between electron and positron by saying that the charged particle that emerges left handed in nuclear beta decay is an electron, and that in our part of the Universe, more of these are present than of its antiparticle. This will only work if it is independently possible to distinguish left-handed from right-handed particles. Such particles cannot be distinguished using strong or electromagnetic interactions, unless a supply of one kind is available, because these interactions satisfy P symmetry. If CP symmetry were exactly true, the weak interactions could not be used either, unless there was an independent way of distinguishing particle from antiparticle, which is what we are trying to accomplish. We could explain what we mean by handedness by transporting some particles of one handedness to the other location. In principle, we could do this by using photons of a definite handedness to communicate, but the usual radio or light beams involve equal mixtures of both types of photons and it would be difficult to avoid this practice.

The decay of K^0_L mesons does afford a simple solution to the problem. Whether the material in a region is composed of what we call matter or antimatter, we can obtain a pure sample of $K^0{}_L$ as follows. We first produce neutral K mesons by some collision process. After waiting about 10^{-7} sec, any $K^0{}_S$ component of the mesons originally produced will have decayed away. The remaining neutral particles of rest energy 498 MeV, whatever their origin, will be $K^0{}_L$. Therefore, in the decay of these particles, a surplus of what we call electrons will be produced. If this procedure is followed in another part of the Universe, whatever type of matter it contains, a surplus of electrons will result. Therefore, a comparison between the lepton that is preferentially produced in $K^0{}_L$ decay with the lepton present in the ordinary matter in that region will determine whether that region contains what we call matter or antimatter. In other words, the decay of $K^0{}_L$, which can be defined by time measurements alone, allows the determination everywhere in the Universe of what is an electron.

The reason for the breakdown of CP symmetry in K^0 decay remains unclear at present. There have been numerous searches for deviations from this symmetry, or the related deviations from time-reversal symmetry in processes not involving K mesons but none have been found. A hypothesis which fits all the data, but casts little light on the reason for it, is that there exists a so-called "superweak" interaction whose strength is much less than that of ordinary weak interactions, and which alone is not CP symmetric. In this model, the effect of the superweak interaction would not show up in places other than K^0 decay using the currently feasible level of accuracy of measurements. The reason that the K^0 system is especially sensitive is that it contains two particles nearly equal in mass, so that any small disturbance can result in a gross distortion of the identity of the particles. No other example of such systems is known where a CP nonsymmetric disturbance could so alter the identities, although similar effects involving other disturbances, such as electric forces, but not CP symmetry, are known in atomic physics. If the "superweak" interaction hypothesis is correct, then a great improvement in experimental technique would be needed to detect the effects of such a small interaction among particles other than the neutral K mesons.

A Theory of Weak Interactions

From the wealth of experimental data on the weak interactions it has been possible to formulate some theories of the fundamental weak interaction processes. The basic assumption made in most such theories is that all of the observed weak interactions take place in two steps, that is, as one virtual transition followed by a second virtual transition, whose over-all result is the observed process. In this respect, the observed weak interactions are taken as analogous to the scattering of two nucleons, or of two electrons, which we have also seen are described as occurring via two virtual processes. In the case of electron–electron scattering the two steps involve the emission and absorption of virtual photons. By analogy, we ask what type of particle can be emitted and absorbed to give the observed weak processes.

One important bit of evidence about the type of particle that may transmit weak interactions is that the weak-interaction processes involving leptons vary in a simple way, as the momenta of the leptons change. For example, the neutrino scattering cross section is directly proportional to the neutrino energy. In contrast the electron-proton or electron-electron scattering cross section is very large at small energies and tends to decrease with energy at a fixed angle. The reason for the latter variation is that the virtual photon that is exchanged has zero mass. According to the relation between the range of interaction and the rest mass of the exchanged particle that is implied by Heisenberg's relation, this in turn allows two charges to interact even when they are very far apart, which results in a large cross section at low energies. The absence of this phenomenon in neutrino scattering, and of corresponding ones in other weak processes, suggests that if a particle is exchanged in weak processes its rest mass is not zero. Indeed, by analyzing the expected variations of cross section in detail, it can be shown that such a particle must have a rest energy of at least 10 GeV, and perhaps a much higher one.

Other inferences that can be made about a hypothetical particle exchanged in the weak interactions concern its spin and charge. Suppose we consider the neutron beta decay. It is natural to imagine that the first step involves the conversion of a neutron to a proton together with the emission of the virtual particle (Fig. 29). The second step would then be the decay of the virtual parti-

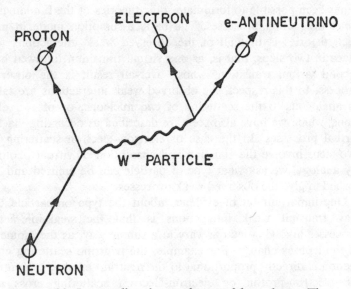

FIGURE 29. Intermediate boson theory of beta decay. The beta decay of a neutron occurs as a two-step process in the intermediate boson theory. A neutron changes into a proton, emitting a virtual W^- particle. The W^- particle travels the stort distance allowed by Heisenberg's relation, and then decays into an electron and an e-antineutrino.

cle into electron and antineutrino. Alternatives would require conversion of a neutron into a lepton, and decay of the virtual particle into proton and lepton; these seem less plausible, considering the great differences between hadrons and leptons. The alternatives also make wrong predictions. Since the virtual particle is produced with a proton from a neutron, it must have charge of $-e$, as well as zero baryon number. These properties allow it to decay into an electron and antineutrino in the second step. We know that the electron and antineutrino which both have spin of $\frac{1}{2}\hbar$, emerge with opposite handedness, which means that when their momenta are also opposite which can happen, they carry away a spin of $1\hbar$. This implies, from the conservation of angular momentum, that the spin of the virtual particle is also $1\hbar$. Because the over-all process of neutron beta decay does not satisfy P symmetry, at least one of the two steps must fail to satisfy it also. Actually, the data on beta decay is consistent with all of the particles in the initial and final state having a preferential handedness, which would imply that neither step is P symmetric.

The hypothetical particle of high mass, spin $1\hbar$, and charge of $-e$ or $+e$ (for the antiparticle) that is emitted and absorbed in weak interactions is called the intermediate vector boson, or the $W^{\pm}$ particle (W for weak, $\pm$ for electric charge). It has not been detected as of this writing, and if its mass is as high as some theories suggest, it will require a new generation of machines that accelerate particles to higher energy than is now possible to produce it. Meanwhile, it is possible to explore the consequences of the hypothesis theoretically in more detail, to see how consistent it is with observed data on weak interactions. The coupling strength of the W particle to leptons and to hadrons is essentially determined by the requirement that its exchange reproduce the observed weak interactions of these particles. The result of detailed analysis is that the interaction of the W particle with hadrons and leptons has essentially the same strength. This suggests that if a W particle could be produced, it would decay into hadrons and leptons with roughly comparable probabilities. The half-life of the W meson depends on its actual rest mass, but cannot be greater than 10^{-21} sec. This is partly because of the high mass the W must have, and partly because its interaction is stronger than the weak interaction. This is known because the W interaction must occur twice to give one weak interaction. Therefore, the probability of a weak interaction is proportional to the square of the probability of a W interaction, and since probabilities are always less than one, the W-interaction probability is larger than the weak-interaction probability. These probabilities are just proportional to the strength of the interaction.

Several attempts have been carried out to produce and detect real W mesons. One of these involves neutrino scattering. A beam of mu-neutrinos is allowed to hit a target of some material. A mu-neutrino can transform virtually into a W^+ particle and a negative muon. One of these virtual particles can then scatter from the target, so that a real W^+ and muon are produced. The need for the second step occurs because a neutrino cannot transform into a real W and muon without violating the conservation of energy and momentum. The scattering of the virtual particle from the target restores the energy that is not conserved in the first step, giving over-all conservation. Of course, both processes must occur within the short time allowed by Heisenberg's relation. The W^+ particle that is produced can decay in several ways, one of which is into a positive muon and a muon neutrino. Since this happens 10^{-21} sec or less, only the decay products can be detected. Therefore, what would be seen in this hypothetical event is the production of a positive and negative muon along with an undetected neutrino and perhaps some hadrons that are produced when the virtual particle scatters from the target. In order for this to be possible at all, the incident neutrino must have enough energy to create the rest mass of the heavy W^+ meson. For neutrinos of the highest available energy, some 400 GeV, a W^+ particle of rest energy no greater than 28 GeV could be produced. The great disparity in energy is caused by the fact that when a high-energy neutrino or other particle collides with a stationary target, most of its energy must go to the velocity of the particles produced; only a small fraction can go to creating the rest mass of new particles. This is not true in colliding beam machines discussed in Chap. VII, but such beams cannot be used with neutral particles. It is believed that existing experiments with neutrinos are sensitive up to no more than 10 GeV W^+ energy, because very few high-energy neutrinos are available. We shall see that according to one theory the rest energy of the W particle must be greater than 37 GeV, and thus it could not be produced by neutrino beams from existing accelerators.

The W-particle model of weak interactions has several interesting consequences concerning new weak-interaction processes. For example, it should be possible for a neutron and proton in a nucleus to exchange a virtual W particle in addition to the usual pion exchange. The W-exchange process would generate a very small force between the neutron and proton, which would differ from the usual nuclear force in that it would not be P symmetric. This would have several experimental consequences, one of which is that in the decay of certain excited nuclear states by photon emission there would be a small preferential emission of photons of one handedness. This is not because of a lack of P symmetry in the electromagnetic interaction that creates the photon, but rather because the nuclear levels themselves are disturbed by the weak force in such a way that even a parity symmetric interaction can produce a preferential handedness. The effect is expected to be only about one part in 10^7 in favorable cases, but it has been observed by very accurate measurements of the photon polarization.

Another effect predicted by this model involves interactions between electrons and e-neutrinos. For example, by having one electron change into a neutrino, emitting a W particle, and another neutrino absorb the W particle, becoming an electron, a kind of scattering between electrons and neutrinos is induced. It has not yet been possible to detect this scattering because the cross sections are very low, even lower than those for neutrino–proton scattering, and because the process itself is difficult to observe even when it occurs—all that can be seen is a moving electron with no obvious origin. However, there may be evidence for related processes such as the emission of neutrino pairs by electrons in hot stars, which in certain cases can have a major effect on the life cycle of the star.

In summary, the W-particle hypothesis seems to provide a reasonable picture of the weak-interaction processes. It is probable that complete conviction about this idea will not come about until the W particle itself can be produced. Meanwhile, various attempts have been made to generalize the hypothesis, and to relate the weak interactions to other known phenomena, especially the electromagnetic interactions.

One such relation that has been explored is the possible existence of neutral counterparts to the W particle. Since the W particle is electrically charged, its creation or annihilation requires a change in the charge of the particle creating it or annihilating it, such as the change of a neutron into a proton. If there were a neutral counterpart to the W particle, it could be emitted without a change in charge of the emitting particle. This would make possible various new weak-interaction processes, such as the scattering of a neutrino by a proton in which the neutrino remains a neutrino and the proton remains a proton (Fig. 30). Such processes have come to be known as neutral current interactions, whereas the usual weak-interaction processes are called charged current interactions. The word "current" here refers to a property of the combination of hadrons that interacts with the W particle, or its neutral counterpart, often called the Z^0 particle. The W and Z^0 particles would have spin $1\,\hbar$, and therefore have a spin direction associated with them even at rest. Therefore, whatever combination of hadrons or leptons they interact with must also have an associated direction in order that the interaction should appear the same to various rotated observers. This direction could involve ei-

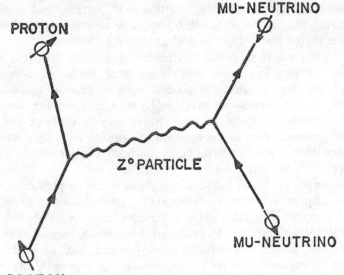

FIGURE 30. Elastic scattering of mu-neutrinos by protons. If neutral as well as charged intermediate bosons exist, certain novel processes can occur in weak interactions. One such process is the elastic scattering of a mu-neutrino by a proton, which has recently been observed in the laboratory. This occurs as a two-step process. The neutrino emits a virtual Z^0 particle, remaining a neutrino. The Z^0 travels a short distance to a nearby proton, where it is absorbed, the proton remaining a proton. Since momentum is exchanged between neutrino and proton, this is observed as a scattering process.

ther the spin of one or more hadrons or the direction of motion of the hadrons. In either case, the direction defined by the combination of hadrons or leptons is described by a mathematical quantity called a current, and therefore the weak interaction is called a current interaction. In the corresponding example of strong interactions involving pion emission, there is no specific direction involved because the pion spin is zero, and thus the interaction is not a current interaction.

In 1973–1974, experiments were carried out to search for neutrino scattering by nucleons that would occur through the exchange of a Z^0 particle which would involve the neutral currents. In these experiments, mu-neutrinos that originate from the decay of high-energy pions were made to strike a target. If the neutrino scattered but remained a neutrino, no charged lepton would emerge. However, if the nucleons in the target converted to other hadrons, they would be observable. Thus the event searched for involved a burst of high-energy hadrons in a target exposed to a neutrino beam. A more detailed description of one such experiment is given below. In the experiments that have been carried out at various laboratories, evidence for the occurrence of such events was found at a rate similar to the rate for events in which the neutrino converts to a muon. It therefore appears likely that neutral current interactions really occur, and therefore that the Z^0 particle is as likely to exist as the W^+ particle.

The existence of a Z^0 particle, and the corresponding neutral current interactions, was predicted by some theories devised in order to solve certain mathematical problems that arose in the theory of weak interactions as mediated by the W particle, and in the theory of electromagnetic interactions of the W particle. These problems occur when one tries to calculate effects in which more than one virtual W particle is exchanged, or when virtual W particles as well as virtual photons are exchanged between hadrons or leptons. Such processes would either give corrections to the simple one-W exchange, or, in some cases, generate new effects that cannot occur via a one-W exchange. In either case, it was necessary to know what theory predicted in order to compare the hypothesis with experimental data. However, an application of the rules of relativistic quantum mechanics to such "higher-order processes" led to ambiguous or infinite answers for quantities that should, according to experiment, be small or zero. This is in contrast to the result of the calculation of similar higher-order processes in the theory known as quantum electrodynamics, which describes the interaction of photons with charged leptons. The predictions of quantum electrodynamics, including higher-order processes, agree with experimental results to great accuracy.

Unified Theories of Weak and Electromagnetic Interactions

In the late 1960s, it was proposed by Steven Weinberg that these difficulties could be eliminated if neutral Z^0 particles existed in addition to W particles, and if the Z^0 particle interactions with hadrons and leptons were related to those of the W particles in a simple way, analogous to the relation between the interactions of charged and neutral pions with nucleons. Furthermore, in certain models of this type, the interactions of W and Z^0 particles were related to those of the photon, which is also a particle with spin $1\,\hbar$. Since the photon interacts at a known strength with charged particles, this model would help determine the strength of the weak interactions. However, the theory is such that the relation between the interactions is simplest at very high energy, whereas for the low energies at which most data on weak interactions exists, the difference between the zero photon mass and the necessarily high masses of W and Z^0 masks the simple relations that exist. Nevertheless, it is possible to infer something about the masses of W and Z from the low-energy weak-interaction data and the known electromagnetic coupling. In the simplest of models, the prediction is that the rest energy of W is at least 37 GeV, while the rest energy of Z is at least 74 GeV. Other models allow lower values of the mass. In most models, the relative strengths of charged current processes and neutral current processes at low energies are also related to the masses of W and Z^0, and so the models make some predictions about this relative strength as well. At present the data on neutral current interactions is not sufficiently detailed to determine whether these predictions are correct.

The theories that relate W interactions, Z^0 interactions, and electromagnetic interactions are generally known as "unified gauge theories." The word "gauge" refers to a mathematical property of the wave functions used to describe the particles in the theory. Several examples of such theories have been proposed with various properties for the particles. In all of these theories, the effect of two or more virtual W particles or Z^0 particles can be calculated, and are found to give small corrections to the single-particle exchange. These results seem to be in agreement with experiments, and thus these theories do solve the problem that they were designed to solve. In principle, it is possible to distinguish between various such theories through experiments, but the relevant experiments have not yet been done. An important feature of all the unified gauge theories is that they explain why the strength of the weak interaction is the same for different leptons and for leptons and hadrons. In certain cases, they also explain the remarkable fact that the electric charges of the subatomic particles are all either zero or integer multiples of a unit charge, which could be chosen as the charge of one type of quark. This latter fact is by no means trivial, because the various conservation laws forbid baryons from converting to leptons, or muon-type leptons from converting to electron-type leptons. A world in which the proton and positron had slightly different electric charges would not contradict laws of physics such as the conservation of charge provided that the charges of some of the other particles were modified accordingly. However, it is known that these two charges can differ by at most one part in 10^{20}, so there is clearly something to be explained.

The kind of explanation for these facts that is offered by the unified gauge theories is analogous to the explanation of another fact: the exact equality of spin angular momentum of different particles such as the electron and proton. The latter explanation is tied to the fact that angular momentum has three components, related to one another by rotations. If the spin angular momenta of different particles did not have integral ratios, it would not be possible for each of the components of angular momentum to be conserved when the particles interact with each other. In some types of unified gauge theories it is assumed that the interaction strengths of W particles, Z particles, and photons with each kind of particle are related to one another by mathematical formulas that are similar to those relating the different components of angular momentum to one another. Therefore, through this assumption the interaction strengths of the different particles are forced to have values that are the ratios of integers or, in some cases, other simple numbers. Through this one mathematically simple assumption these theories are able to account for several facts at once.

At present, the unified gauge theories appear to be the most satisfactory description of weak interactions. The qualitative prediction of neutral current effects made by these theories has apparently been verified. More convincing evidence for these theories would be the discovery of the W and Z particles and the obtaining of quantative results for the various effects predicted by these theories. A good deal of high-energy physics research in the coming decade is likely to be devoted to these questions. An important theoretical question about weak interactions that remains is the origin of CP noninvariance, which does not appear to have a very natural place within the unified gauge theories, although several models have been constructed in which the two are combined. These latter models tend not to explain CP breakdown as due to a superweak interaction, but rather to assume that the amount of breakdown is accidentally small in the low-energy processes that have been studied. Information about the validity of CP invariance in high-energy weak-interaction processes would therefore be important for deciding whether these models are reasonable.

Finally, we may ask how these ideas about weak interactions are influenced by the quark model of hadrons. In some respects, it is simple to fit the two together. We can imagine that the W and Z particles interact directly with the quarks that constitute the known hadrons, just as we imagine that photons interact directly with the charged particles that make up atoms. The weak interaction processes that are observed to occur between hadrons can then be interpreted as transformations among the quarks. From this viewpoint, the weak interaction decays of ordinary hadrons would be analogous to the beta decay of a nucleus, which is now interpreted as a transformation of one of the nucleons contained in the nucleus, rather than a process involving the nucleus as a whole. For example, the pion decay into muon and neutrino can be pictured as the annihilation, into a virtual W meson of the quark–antiquark pair making up a pion, followed by the decay of W into a muon and neutrino. In order to describe processes such as decay of strange particles, it is necessary to allow the weak interactions to break the conservation laws of the number of each type of quark which are satisfied by strong interactions. This can be done simply by allowing a quark of type one to change into either a quark of type two or of type three by emitting or absorbing a W particle. The ratio of the strengths of the two types of transitions must be obtained from experiment. Because the weak interactions in which hypercharge changes are small compared to those in which it is conserved, this ratio turns out to about $\frac{1}{5}$. The reason for the nonconservation of quark types in weak interactions and for the particular number that characterizes this nonconservation are unknown at present.

In contrast to the W-particle interactions, it is essential that the interactions of the Z particle conserve the number of each type of quark. If this were not the case, then certain neutral current decay processes would be observed; in fact they are not. For example, a charged K meson consisting of a type-one quark and type-three antiquark could decay into a charged pion together with a pair of neutrinos. This would happen by the conversion of a type-three antiquark in the K meson into a type-two antiquark together with the emission of a virtual Z particle. The type-two antiquark and the remaining type-one quark could form a charged pion. The Z particle could then decay into a neutrino-antineutrino pair completing the process mentioned. However, if this process occurs at all, its rate is less than 10^{-5} that of a process such as K decay into muon and neutrino, which involves corresponding interactions of the W particle. The conclusion is inescapably that some definite principle is at work in the Z^0 interactions which suppresses the change of type-two quarks into type-three quarks. Indeed, the rate of incidence for the process described above is so low that it is even smaller than what one would expect from the emission and absorption of two W particles, which necessarily can induce the change of quarks required to make this process happen. This latter process could occur by having a type-three antiquark convert into a virtual type-one antiquark and a W^+ particle, followed by the conversion of the type-one antiquark into a type-three antiquark and a W^- particle. The W^- can then decay into an anti-

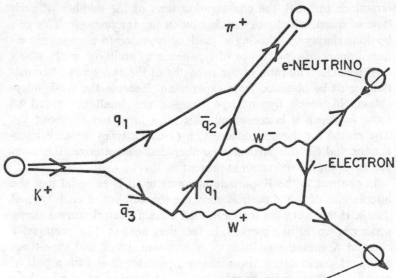

FIGURE 31. Weak-interaction process involving two virtual W bosons. Certain decays cannot occur through a single virtual intermediate boson, but can occur through two or more bosons. These are called higher-order processes, and one such is shown here. A K^+ meson is made of q_1 and a $\bar{q}_3$. The $\bar{q}_3$ can convert into a $\bar{q}_1$ by emitting a W^+ particle. The $\bar{q}_1$ can then emit a W^-, becoming a $\bar{q}_2$, which combines with the original q_1 to become a positive pion. The virtual W^+ and W^- exist for the short time allowed by Heisenberg's relation, and then convert, in two steps, into a neutrino and antineutrino. If this process were observed, it would be an indication that such multiboson effects really exist in weak interactions.

neutrino and an electron, and the electron can absorb the W^+, becoming a neutrino (Fig. 31). The estimated rate for this process in the unified gauge theories is about 10^{-4} of the K^+ decay, or ten times larger than the experimental limit. These higher-order processes are not just theoretical constructs. The very small mass difference between K^0_L and K^0_S is believed to originate just from such processes. Therefore, some mechanism must be found that will suppress the higher-order processes that could change type-two quarks into type-three quarks without eliminating all higher-order processes.

A mechanism that has been proposed to do this involves the fourth quark, known as a "charmed" quark. According to this idea, it is possible to get a conversion of a type-two quark to a type-three quark in two ways. One is the higher-order process described above involving a virtual quark of type one. The other would be a similar process involving a virtual quark of the new type. In order for this to happen, there must be weak interactions involving the "charmed" quark with the type-two quark and with the type-three quark, together with the W particle. By a suitable choice of the strength of these interactions, it is possible to arrange a cancellation between the two higher-order processes, and so suppress the unobserved process. This suppression would be complete if the "charmed" quark had the same mass as the type-one quark. However, since hadrons containing the "charmed" quark are substantially heavier than those involving the other quarks, the "charmed" quark must have a different mass. Thus the suppression is not perfect, and, decays such as K^+ decay into pion, neutrino, and antineutrino should occur at some small rate.

If the "charmed" quark theory is correct, it should be possible to produce hadrons containing this fourth quark by weak interactions such as in neutrino scattering experiments. These "charmed" hadrons would transform into ordinary hadrons only by weak interactions, and so would have relatively long lifetimes. It is expected that their rest energies would be several GeV. Experiments to detect such particles are being conducted at several laboratories, and some evidence for them has been found. If this result stands up, the quark model will have passed its most severe tests in connection with the weak interactions.

A Weak-Interaction Experiment

In order to give the reader some idea of the techniques and problems in experimental high-energy physics, I shall describe in detail the first experiment in which neutral current interactions of neutrinos with hadrons were observed. The experiment was carried out at the CERN laboratory in Geneva, Switzerland. It is typical of many contemporary high-energy physics experiments in that some fifty-five physicists, from at least seven countries, signed the article announcing the result of the experiment. Many more technicians were probably involved as well.

An over-all view of the experimental setup is given in Fig. 32. It begins with protons that are accelerated to an energy of about 26 GeV in the CERN proton synchrotron. These protons are extracted from the synchrotron by magnets into an external beam containing approximately 10^{12} protons in each pulse, with one pulse being produced every 2.5 sec. The external beam is allowed to hit a target made of beryllium, about 30 cm long. This is enough beryllium to ensure that most of the protons will interact at least once in passing through the target. The result of these interactions, as we have learned in Chap. VIII, is that in each proton–nucleus collision, several pions will be produced on the average. Because of the high energy of the incident proton, and because the pions produced tend to have low momenta perpendicular to the beam, the pions continue in approximately the same direction as the incident proton beam, with an average angular spread of only a few degrees.

The charged pions can be focussed by magnetic forces so that they travel down a tunnel some 70 m long. By choosing the magnetic force, it is possible to focus either positive or negative pions selectively. The neutral pions cannot be focussed, and anyway decay too quickly to be of use. The charged pions are travelling at about 0.9999 times the speed of light. At this speed, the time slowdown effect increases their half-life to about 10^{-6} sec from its value of 2×10^{-8} sec for pions at rest. In this time the pions can

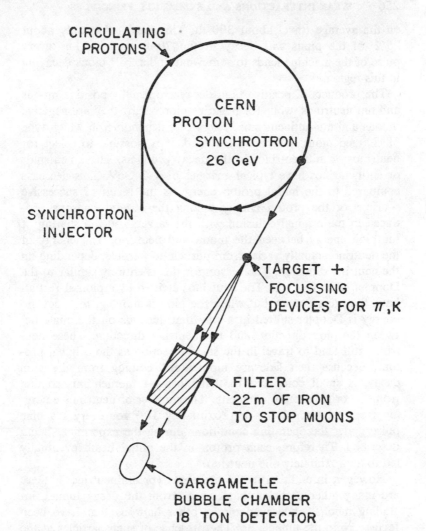

FIGURE 32. Floor plan of the CERN neutral current experiment. The layout of the CERN neutral current experiment. Protons are first accelerated to 26 GeV in a synchrotron, and ejected from it. They hit a target, producing pions and K mesons. These decay in a long pipe, giving neutrinos. A long, iron "filter" removes charged particles that occur in these decays, leaving only neutrinos, and some neutrons. The neutrinos then hit the Gargamelle bubble chamber, where some interact and are detected.

on the average travel about 300 m. This means that only about 25% of the pions will decay in the 70-m tunnel. Thus every pulse of the machine leads to somewhat under 10^{12} pions decaying in this region.

The products of positive pion decay are usually positive muons and mu-neutrinos, while for negative pion decay, they are negative muons and mu-antineutrinos. Therefore, depending on which type of charged pions have been focussed, it is possible to select for neutrinos or antineutrinos in the decay products. These neutrinos or antineutrinos have typical energies of 1–2 GeV. This decrease compared to the initial proton energy is the result of successive sharings of the proton energy among the several hadrons produced in the original collision with the target, and the sharing of the pion energy between the muon and neutrino. The energy of the neutrino actually varies from particle to particle, depending on the neutrino direction. The reason for this is entirely similar to the Doppler effect for light. The neutrinos emitted by a pion at rest all have the same energy, but when the pion is moving, the neutrino energy is Doppler shifted in a way that depends on the angle between the pion direction and the neutrino direction. These neutrinos still tend to travel in the same direction as the original protons, because their sidewise momentum, coming from the pion decay, is small compared to their forward momentum, coming from the original proton velocity. The number of neutrinos emerging from the tunnel was approximately 10^{12} for every machine pulse under the operating conditions during the experiment being described. Therefore, each proton in the initial beam eventually led to approximately one neutrino.

However, in addition to the neutrinos (or antineutrinos), there are many other particles that emerge from the decay tunnel, including muons, undecayed pions, other hadrons that have been focussed into the tunnel, etc. The presence of such particles at the actual neutrino detector would be very unfortunate, because all particles other than neutrinos have much greater interaction cross sections than neutrinos do. That is, hadrons could give strong interactions in the detector, muons or photons could give electromagnetic interactions, etc., and all of these are much more probable than the weak interactions of the neutrinos. This would make it very difficult to pick out the neutrino interactions from the background of uninteresting interactions. In order to avoid this, a large amount of shielding is introduced between the end of the tunnel and the detector. This shielding consists of 22 m of iron, which is enough to absorb or deflect almost all particles other than neutrinos, but has almost no effect on the flux of neutrinos, due to the small interaction probability. Actually, in the experiment being described, it was found that a number of neutrons with energy less than 1 GeV did manage to penetrate the shield, but no higher-energy ones appeared to do so. As a result, the experiment concentrated on interactions of neutrinos of energy greater than 1 GeV, in order to avoid confusion with the interactions induced by the neutrons.

The detector used in this experiment, which is placed beyond the iron shield, was a large bubble chamber, given the name "Gargamelle," after the mother of Gargantua in Rabelais' novel. The bubble chamber was filled with eighteen tons of freon, a liquid compound of carbon, fluorine, and bromine used as a heat-transfer agent in refrigerators. The purpose of using freon was twofold. The high density of freon compared to hydrogen increased the number of nucleons in the effective target for the neutrinos, which was the material in the bubble chamber. Second, the high density and high nuclear charge of the atoms in freon mean that any hadron, electron, or photon produced in the neutrino interaction will interact within the bubble chamber by methods other than simple ionization. On the other hand, muons will interact only by ionization, since they have no strong interactions, and the probability of radiation in collisions is much less for muons than for electrons. Therefore, muons can be distinguished from other particles produced in Gargamelle by their lack of observable interactions. A magnetic force exists over the whole volume of the bubble chamber, making it possible to distinguish positive from negative particles, and to measure the momentum of such particles.

The useful volume of the bubble chamber in the experiment was a cylinder 3.75 m long and about 1 m square. Approximately 5×10^9 neutrinos entered this region during each pulse of the machine, and the region of interest was photographed from many angles at each pulse. A total of 83,000 pulses in which the particles were predominantly neutrinos, and 207,000 pulses in which the particles were predominantly antineutrinos were used. In most of the pictures taken, nothing at all was seen, indicating that the shielding was successful in removing almost all of the unwanted particles. It is furthermore not surprising that few pictures showed neutrino interactions. To understand this, we use the fact known from earlier neutrino experiments that the interaction cross section for neutrinos with energy of a few GeV is approximately 10^{-38} cm^2. Since the spacing of the atoms in the freon is approximately 10^{-8} cm, this implies that each neutrino has approximately one chance in 10^{22} of interacting with each atom that it encounters. In passing through the 3.75 m length of the useful volume of the bubble chamber, each neutrino encounters about 4×10^{10} atoms. Therefore, each neutrino has about one chance in 3×10^{11} of interacting in the bubble chamber, and the 5×10^9 neutrinos in each pulse will result in one interaction in every one hundred pulses. Therefore, only 1% of the pictures should have shown neutrino interactions, and this is approximately what was seen. Actually, there were about five hundred pictures showing neutrino interactions and about two hundred showing antineutrino interactions.

These pictures were examined first by trained technicians with equipment that enabled them to identify and measure each track in the picture. The measurements for each picture were fed immediately into a computer, which "reconstructed" the event, i.e., determined which charged particles were involved from an estimate of their mass, and whether any neutral particles must be present also to conserve energy and momentum. Two general types of pictures were found, corresponding to distinct interaction processes. In one type of picture, a long track occurred, showing no interactions (other than ionization) in the bubble chamber. This was the type of track identified with muons. These pictures may also contain one or more tracks originating at the same point as the muon track. These are identified as hadrons (Plate 8). These events were then identified as a neutrino interacting with a proton or neutron in the freon, producing a muon, and either breaking up the nucleus, or making extra pions or other hadrons. These interactions are the previously known charged current interactions of neutrinos. Approximately 80% of the pictures seen were of this type.

The other 20% of the pictures showing interactions—amounting to about 166 in all—were different in that there was no long muon track. Instead, only hadron tracks were seen, in which the visible hadrons, i.e., charged hadrons, or neutral particles that converted into photons in the chamber, have a total energy of more than 1 GeV (Plate 9). These events were the candidates for the neutral current interactions of the neutrinos. It is imagined that a neutrino hits a nucleon in the freon, remains a neutrino, but transfers some of its energy to the nucleon, or produces some additional hadrons, and these moving hadrons are what are seen in the picture. Of course, the incoming and the outgoing neutrino remain invisible as they do not ionize, and they have a negligible chance of interacting more than once in the chamber. The relation between the charged current events and the neutral current events may be clarified by the following analogy. Imagine that an ordinarily dressed thief enters a house, and steals some money. In addition, he steals a flaming red scarf which he puts on and wears on his way out. The scarf makes it relatively easy to detect him afterward on the street. On the other hand, if the thief simply took the money, leaving the scarf behind, he would be difficult to detect once he escaped from the house, and only the missing money indicates that a crime has been committed. The thief with the red scarf is like the neutrino that has changed into a muon, while the thief without the scarf is like the neutrino that remains a neutrino.

The problem with the muonless events is that there are conceivably other interactions that could give similar pictures: a neutron scattered from a nucleus, for example, would do this. The experimenters had to go through a detailed consideration of such other possibilities before convincing themselves that most of the muonless events they saw were really due to neutrino interactions. Even then some physicists remained skeptical, and only after the results of the CERN study were duplicated by several other experimental groups using neutrinos of different energies, and using different detecting equipment, did all physicists become convinced that the neutral current interactions of neutrinos had been found.

XI

The Achievements and the Future of Twentieth Century Physics

Our survey of atomic and subatomic physics has brought us to the questions that physicists are presently trying to answer. To go beyond that point involves speculation that is much less solidly grounded than the ideas discussed previously. Before attempting such speculation, it is useful to summarize the broad trends that have occurred, on the assumption that some continuity is likely to persist in the future.

At the start of the twentieth century, physicists had just begun to probe the phenomena associated with individual atoms, objects with a size of about 10^{-8} cm. In the seventy-five years that have elapsed, we have successively discovered, and understood to various extents, subatomic phenomena down to a level of size of about 10^{-14} cm—a million times smaller than that of atoms. The ratio of these sizes is about that between the Earth and a house, or that between the smallest object visible to the eye (10^{-2} cm) and an atom.

In the course of these discoveries, we have not only been able to understand most of the subatomic phenomena themselves, but have also been able to use the properties of the subatomic particles to explain many larger-scale phenomena that were previously mysterious. These include the electrical properties of materials, the existence of permanent magnets, the origin of the Sun's energy, and the abundance in the Universe, and chemical properties of the elements. While there remain some aspects of larger-scale phenomena that are not yet understood in detail, or quantitatively, in terms of atoms and their constituents, such as the set of processes that we call life, nevertheless, many scientists are convinced that the essential features of all such phenomena are contained in the laws of atomic and subatomic physics already known to us. To a large extent, this belief has been confirmed by the gradual extension of explanations using the concepts of physics into areas such as molecular biology.

The progress we have made in understanding atomic and subatomic phenomena has been possible partly because of two major developments in theoretical physics: the quantum theory and the special theory of relativity. Interestingly, ncither of these theories was originally devised for the purpose of describing subatomic phenomena. The quantum theory was first applied to the emission and absorption of radiation by matter, without reference to its atomic structure, and the relativity theory to the behavior of rapidly moving macroscopic objects. However, it gradually became clear that the essential aspects of atomic and subatomic processes could be understood by a systematic application of quantum theory and relativity theory. Both of these theories involve gross departures from the concepts that were previously used to describe physical phenomena, and it must be considered a triumph of the human imagination that we were able to arrive at ideas that are so foreign to everyday experience and yet describe a wide range of phenomena accurately.

There is an interesting sidelight to the development of twentieth century physics. Perhaps as a consequence of the success of quantum theory and relativity theory and of the accuracy of some of their more surprising predictions, such as the creation of particles, there has been a gradual shift in the balance between theoretical and experimental physics. At the beginning of the twentieth century, purely theoretical studies were relatively uncommon, and many of the studies that were conducted were responses to some definite observations. For example, Planck's original proposal of the quantum nature of light was an attempt to explain experimental data on emission of radiation by hot bodies. However, as the century has progressed, there has been more and more purely theoretical work which attempts to predict the results or to suggest the direction of future experimental work. Indeed, much work in contemporary theoretical physics is not directly intelligible to many experimental physicists without some further interpretation. While it is still true that experimental discoveries unanticipated by theorists are sometimes made, and require significant revisions in existing theories, much of the experimental work in subatomic physics today is done in direct response to theoretical suggestions, rather than conversely. Indeed, in many cases it is difficult for experimental physicists to obtain access to the complicated equipment they need for their experiments, unless their proposal makes some reference to a theoretical idea for which these experiments are relevant. In view of the historical development of physics as an effort to understand observations and experiments, this situation is perhaps not satisfactory. It remains to be seen whether the present hegemony of theory over experiment will continue, and what its ultimate effect on the progress of physics will be.

While physicists have been fairly successful at explaining most of the subatomic phenomena they have uncovered, we have not yet reached any level in which the phenomena themselves are much simpler than those on higher levels. In that respect, subatomic physics has not yet fulfilled one of the early hopes that it would eventually lead us to the ultimate constituents of the Universe, and that these constituents would require only a few simple ideas for their description. Instead, while the successive levels of atoms, nuclei and electrons, and hadrons and leptons have each helped to explain the properties of the next-higher-level objects that they compose, the phenomena within each of these new levels are rich and varied, and the properties of the objects on each new level are themselves complex. Perhaps some extension of the quark model represents the simplicity we are seeking, but already the properties assigned to the quarks, in order to describe the observations made on hadrons, make the quarks seem relatively complicated to be such ultimate constituents.

The various levels of subatomic phenomena we have uncovered

thus far have all been understood in terms of particles, that is, small discrete concentrations of mass and other properties which move through space. While this notion of particle is consistent with the idea of relativity and quantum mechanics, it is not the only type of object that is so consistent. Furthermore, while relativistic quantum theory places many restrictions on the type of objects that can exist in the Universe, it does not uniquely determine which objects do exist and which do not. There have been some theoretical speculations that have attempted to fit certain of the phenomena involving hadrons into a description in which the fundamental objects are not particles, but instead are some of the other possibilities allowed by relativistic quantum theory. Such objects, known as solitons, differ from particles mainly in that they have a finite intrinsic size. Efforts to use such objects to explain phenomena are complicated by the fact that these alternative possibilities have not been completely analyzed, so that their precise properties are not even known theoretically. At present, these ideas are expressed in a highly mathematical form, and it is far from clear what kind of intuitive picture can be extracted from them. If solitons do turn out to give a correct description of hadron phenomena, it will be worthwhile to see if such an intuitive picture can be found, and to decide what new features beyond those of particles are included in it.

In any case, it does not appear at the present that we have reached a complete and simple understanding of the phenomena of subatomic physics. It is conceivable such an understanding will require the discovery of new phenomena, which will help to clarify those already known, or perhaps give an indication of some deeper levels of structure of subatomic objects. It is of course very difficult to make reasonable predictions about what new phenomena may be discovered, and about possible explanations for them. However, I offer the following remarks in a highly speculative vein that I have tried to avoid in the rest of this book.

One obvious uncertainty is whether the present description of subatomic processes by relativistic quantum theory will remain adequate for any new phenomena discovered in the future. While this description has been successful over a wide range of sizes and energies, it may have its limitations, just as other physical theories have had. This might show up either as a small deviation from the predictions of relativistic quantum theory for already known phenomena, or by the discovery of new phenomena which deviate grossly from these predictions. As an example of the latter, we might find objects whose angular momentum is not an integral or half-integral multiple of $\hbar$ as required by relativistic quantum theory.

At the moment, the only indication of a limitation on the applicability of relativistic quantum theory is for phenomena involving gravitation. According to Einstein's general relativity theory, gravitational phenomena are known to produce changes in the properties of space and time that cannot be described within the special theory of relativity. A synthesis between general relativity theory and quantum theory has not yet been completely achieved, although some progress along these lines has been made. Gravitational effects are believed to be very small on the scale of sizes and energies that has been explored thus far in subatomic physics, and are probably indetectable. However, at energies of 10^{19} GeV, or for hypothetical objects whose size is 10^{-33} cm or less, the gravitational effects would become comparable to those of electromagnetic and strong interactions. While we cannot yet tell what would happen at such energies or sizes, there remains a very wide range of intermediate values that have not yet been explored in which gravitation probably plays a minor role, and in which the validity of special relativity and quantum mechanics remains an open question. The study of phenomena in that range of energies and sizes may have some surprises for us that could require a new description of physical phenomena.

An obvious development within the present framework that might be anticipated is the discovery of new particles. Almost certainly, many new hadron states of the types already known will be found. Most of these will be unstable, but some metastable ones such as more bound states of ordinary quarks and new may also be discovered. Perhaps more interesting is the possibility of discovering new particles that are different in some important quality from those presently known. A number of speculations about such particles have been given by theoretical physicists over the years, and some such as the quarks themselves and the W and Z particles were discussed in previous chapters. Discovery of these would be important, if somewhat prosaic, confirmations of theories that attempt to explain properties of known particles. There are also other proposed particles that are not so intimately connected to known phenomena.

One such type of particle is known as the magnetic monopole, a particle that would generate magnetic forces in a way similar to the way in which electric charges generate electric forces. That is, a monopole at rest would exert a magnetic force on another monopole at rest. We of course know that magnetic forces exist, but they are produced by moving electric charges. An electric charge at rest, without spin, will produce no magnetic forces. While magnetic monopoles were thought to exist in the eighteenth century, the discovery by Oersted that an electric current produces magnetic forces made them appear unnecessary, and none were ever discovered. However, around 1930 it was pointed out by Dirac that if such magnetic charges nevertheless existed, it would help account for a fact that is not completely understood, i.e., that all known particles have either zero electric charge, or an integer multiple of the electron charge. (If quarks exist, this would have to be changed to an integer multiple of the quarks' charge.) The relation between the existence of monopoles and this "quantization" of charge is not easy to demonstrate without mathematics, but the arguments for it are convincing. This does not mean that the quantization of electric charges requires the existence of monopoles. Rather, the existence of monopoles would be one way to account for the quantization. As a result, there have been several searches for monopoles in the last forty years, with accelerators, in cosmic rays, and even on the Moon. One may have been found in 1975, but the evidence is questionable. Since Dirac's argument and subsequent investigations have not predicted any definite rest mass for monopoles, it may be that they exist, and that we have just not yet been able to produce them with existing accelerators.

Another new type of particle called the *tachyon* has been hypothesized by Gerald Feinberg, and independently, by George Sudarshan and co-workers. This is a particle whose speed in vacuum would always be greater than, rather than always less than, the speed of light. Such a particle, when created, would be traveling faster than light, and would continue to do so through whatever interactions it had until it was annihilated. The existence of tachyons would not contradict the special theory of relativity, which forbids any particle from making a transition across the "light barrier," but does not forbid particles that always remain on either side of the barrier. Tachyons would have several strange properties. For example, they would speed up as they lose energy, say by collisions with atoms, until as their energy approached zero, they would approach infinite speed, and so be at all points on a line at the same time! The existence of tachyons would not solve any known problems, and they can be regarded as an example of a possibility that is allowed by present day theoretical physics without being required by it. Several experimental searches for tachyons have been made, but no evidence for them has been found, and it appears that if they exist at all, their interaction with ordinary forms of matter must be extremely small.

Still another possibility for new types of particles would be other examples in the family of particles called leptons. We have seen that the discovery of the muon was made easier by the fact that pions decay into muons. If there were other leptons, that is, particles without strong interactions, much more massive than known leptons, it would be much harder to produce them. If such heavy leptons have electric charge, they might be produced in pairs by the annihilation of high-energy electrons and positrons. Some evidence for heavy leptons was reported in 1975, from the Stanford Linear Accelerator Laboratory. The discovery of new heavy leptons might make possible a more accurate classification of this family in analogy to those that have proven successful for the hadrons.

Another plausible question, based on what has gone before in subatomic physics, is whether any new types of interaction among particles will be discovered. In order to be really new, rather than just another aspect of known interactions, an interaction would have to have some qualitatively different properties. For example, it might not satisfy some of the conservation laws that are satisfied by known interactions. If this is the case, it is not possible to generate the new interaction from real or virtual effects of the old ones. One candidate for this type of new interaction has been mentioned, the so-called "superweak" interaction that violates CP invariance. If this interaction exists, it is much weaker than the weak interaction. This illustrates a phenomenon that has been noted by several physicists; the weaker an interaction is, the more symmetry principles it seems to violate. If superweak interactions exist, they may also violate other symmetry principles such as electron number conservation or muon number conservation. It has even been proposed that baryon number conservation is not exactly satisfied, and that there exist extremely weak interactions that can lead, for example, to the instability of the proton. If such interactions do exist, they must be 10^{50} times smaller than even the superweak interactions, because of the known upper limit of 10^{28} years for the proton lifetime.

Another way an interaction might qualify as new would be to involve combinations of particles that are not presently known to interact. This is harder to make precise, because any two particles may be expected to interact through some virtual effects of known interactions. However, in some cases this interaction is likely to be quite small, and so there is some room for a search for a new interaction between certain particles that is larger than these virtual effects. For example, neutrinos and photons are expected to interact via the virtual effects of weak and electromagnetic interactions. This interaction could lead to the production of neutrino–antineutrino pairs in certain hot stars from the photons present in the star. It is possible that neutrinos and photons have an additional direct interaction that is much larger than this indirectly caused interaction, and if so, this could be discovered through astronomical measurements.

Finally, it is possible that there may be undiscovered interactions that differ qualitatively from known interactions through their dependence on the distance between the interacting particles, or on the number of interacting particles. An example of this would be the hypothetical interaction that keeps quarks confined inside hadrons, an interaction which increases rather than decreases with distance. Another possibility is an interaction that requires the presence of more than two particles near one another in order to be large enough to be measurable. This also has a precedent in the case of gravity, which is unobservably small between single subatomic particles, but becomes the dominant interaction when astronomical numbers of particles interact with one another. It is imaginable that other such interactions exist which are relatively unimportant for one or two particles, but become significant when one hundred or one thousand particles are near one another. Such interactions might play a role in the properties of heavy atoms or nuclei, but no evidence for them has surfaced as yet.

In view of the many possibilities for novel interactions, it would be surprising if we did not discover some qualitatively new interactions in the future of subatomic physics.

A different type of question is whether we will develop new experimental techniques for dealing with subatomic particles. Because of the increasing size and cost of the equipment presently used in this research, it seems clear that such new departures are necessary if the research is to continue successfully. Some different approaches to the acceleration of particles to high energy have been proposed, although none are yet in use. One of these involves the use of the very intense electric and magnetic forces that can be produced in and near a stream of charged particles. The charged particles producing these forces need not have a very high energy, but the forces, acting on other charges that are introduced into the stream, can under some conditions accelerate them to extremely high energies. These energies could only be obtained —at prohibitive expense—in conventional accelerators by the use of immense externally produced electric and magnetic forces. The idea involved here is vaguely similar to that of the laser, i.e., relatively small amounts of energy are given to many different particles, and the mutual interaction of these particles converts this into a large amount of energy for a few particles. This would be a step in the direction of using specific properties of the subatomic particles to control their behavior, instead of the present approach of using the bulk properties of matter to do so.

However, it does not at present appear likely that we shall ever be able to control the behavior of individual subatomic particles (or atoms, for that matter) even within the limits allowed by Heisenberg's relation. The disparity in durations, sizes, and masses between subatomic processes and those of our instruments appears too great to be overcome in any foreseeable way. In this regard, the physicist is in an even worse position with respect to subatomic particles than is a bacteriologist to the bacteria that he studies. This is not very important from the viewpoint of understanding the particles, because all particles of a type are identical, so that dealing with them as individuals is no better than making statistical inferences from the behavior of many of them. However, it does cast doubt on the prospects that we will ever be able to make practical use of most of the properties we have found the subatomic particles to have.

As we probe more and more deeply into the structure of matter, we are probing levels of size, duration, and mass that are further and further removed from those of our bodies, or the instruments that are extensions of them. It is not surprising that as we do this, it should become harder and harder to deal with the objects we find there, except through the use of highly abstract interpretations of what we observe. A similar process has occurred on the opposite end of the scale, in astronomy, dealing with objects the size of the visible universe, or the density of neutron stars. We do not know how much further we shall have to probe into subatomic phenomena before we reach an end to novelties, if indeed that will ever happen. Nor do we know if we as individuals, and as a species, are capable of scientific investigations to the point where this happens. These are questions for future humans to answer. For the present, we can be proud of our accomplishments in first unveiling, and then understanding that part of the subatomic world we now know. It is my belief that this will stand as the major intellectual accomplishment of mankind in the twentieth century.

Glossary

ALPHA DECAY

A type of spontaneous nuclear transformation in which a nucleus breaks into two parts, one of which is a helium nucleus, or ALPHA PARTICLE. Several naturally occurring heavy nuclei are radioactive through alpha decay, which led to the original discovery of radioactivity.

ANGULAR MOMENTUM

A quality associated with particles that either are moving in a curved path or are rotating on an axis. The former type is called orbital angular momentum, and the latter is called spin. Angular momentum has both a direction and an amount. Both of these can assume only certain values according to quantum mechanics.

ANTIPARTICLES

Two types of particles with equal mass and spin but opposite electric charge, baryon number, and other additive properties are known as antiparticles. Examples are electrons and positrons, or positive and negative pions. Each of the two types of particle can be thought of as the antiparticle of the other and neither is more fundamental, although one type is sometimes more plentiful on Earth. Particles without additive properties, such as photons, are identical with their antiparticles.

ATOM

The smallest unit in which a chemical element can occur. Atoms consist of a positively charged nucleus containing most of the mass and a number of negative electrons moving at some distance from the nucleus.

BARYON

Any one of the hadrons with a spin that is a half-integer multiple of $\hbar$. Examples include protons and neutrons. All baryons carry a conserved quantity known as baryon number which restricts the way the baryons can transform or decay.

BETA DECAY

A type of nuclear transformation involving weak interactions. In the most common kind of beta decay, a neutron, either free or bound in a nucleus, converts into a proton while emitting an electron and an antineutrino. In another form, a proton in a nucleus changes into a neutron, emitting a positron and a neutrino.

BOSON

A member of a class of particles with the property that any number of each type can be found in the same state at one time. All particles, such as pions or photons, whose spin is an integer multiple of $\hbar$ are bosons.

BUBBLE CHAMBER

A device used to detect rapidly moving, charged subatomic particles. It is a container filled with liquid at the point of boiling. A particle passing through the liquid causes bubbles to form along its path. These bubbles can be photographed, providing a record of the particle's motion.

CONSERVATION OF ENERGY

A law in Newtonian mechanics and in quantum mechanics. It states that the sum of the rest energies, kinetic energies, and potential energies of a set of interacting objects remains constant in time, although the individual quantities may change. In quantum mechanics, the law is subject to the qualification that it may be violated for short periods of time, because of Heisenberg's relation between energy and time.

CONSERVATION OF MOMENTUM

A law in Newtonian mechanics and in quantum mechanics which states that the sum of the momenta of any set of interacting particles does not change with time either in its amount or in its direction, provided that no forces from outside the set act on the particles.

CROSS SECTION

The probability that a scattering process will take place when two particles approach one another is measured by a number called the scattering cross section. The cross section represents the effective target area displayed by one particle to the other. It is therefore measured in units such as cm^2.

DECAY

A transformation in which an atom, nucleus, or subatomic particle changes into two or more such objects, whose total rest energy is less than the rest energy of the original object.

DEGENERATE STATES

Two or more quantum-mechanical states are said to be degenerate if they have the same energy, but differ in some other quantity, such as angular momentum. The total number of degenerate states corresponding to one value of energy is called the degeneracy.

DIFFRACTION

When a wave passes by an object with an edge, there is not a perfectly sharp shadow of the object produced. Instead, some of the wave will be bent by the edge into the region of the shadow. This bending is called diffraction. The amount of diffraction is greatest for waves of longest wavelength.

DOPPLER EFFECT

The energy of a photon, or equivalently, the frequency of a light wave increases when the source and the observer of the light move toward each other, and decreases when the two move away from each other. This variation in frequency is known as the Doppler effect, and is a consequence of relativity theory.

e

The unit of electric charge. The charge of a proton is $+e$, that of an electron is $-e$. All particles observed until now have charges that are integer multiples of e.

ELECTROMAGNETIC INTERACTION

The process involved in the emission or absorption of photons by charged particles. Electric and magnetic forces between charges result from the repeated occurrence of electromagnetic interactions involving virtual photons.

ELECTRON

The least massive particle carrying electric charge, and a constituent of all ordinary atoms. Electrons have spin of $\frac{1}{2}\hbar$, and are stable against spontaneous decay.

ELECTRON VOLT (eV)

A unit of energy in atomic and subatomic physics. One electron volt is the energy that would be obtained by a body with a charge equal to that of an electron moving across an electric potential difference of 1 V. There are approximately 2×10^{25} eV in 1 kWh.

EXCLUSION PRINCIPLE (PAULI PRINCIPLE)

A physical law restricting the behavior of certain particles, such as electrons. For these particles, called fermions, it implies that no two particles of any one type can be found in the same quantum state at any time.

FERMION

Any member of the class of particles obeying the exclusion principle. No two fermions of the same type, such as electrons, can be found in the same state at one time. All particles, such as electrons, protons, and neutrons, whose spin is $\frac{1}{2} \hbar$, are fermions.

FREQUENCY

The frequency of a wave that repeats its form periodically and moves through space is the number of wave crests that pass any point in space in some interval of time.

GAMMA RAY (γ RAY)

A beam of high-energy photons of the type emitted in some nuclear or subatomic particle decays. Gamma rays are similar to X rays in much of their behavior, and the distinction between the two is largely a matter of convention.

$\hbar$ (AITCH BAR)

A number important in quantum mechanics because it indicates the extent to which momentum and position can be simultaneously determined and also because it is the natural unit for orbital angular momentum. $\hbar$ is equal to approximately 10^{-27} g cm^2/sec. Planck's constant is $2 \pi \hbar$.

HADRONS

Any one of the many types of particles that undergo strong interactions. Hadrons are easily produced and easily transformed into one another, provided that enough energy is available.

HALF-LIFE

The length of time in which there is a 50% probability that a certain decay process will occur is called the half-life for that decay. Each individual atom or subatomic particle may decay at any unpredictable time. However, in a sample containing many atoms approximately 50% of the decays will occur within one half-life.

HEISENBERG'S RELATION (UNCERTAINTY RELATION)

A fundamental principle of quantum mechanics restricting the extent to which momentum and position can be simultaneously determined for any object. It states that the product of the extent to which momentum is determined and the extent to which position is determined cannot be smaller than $\frac{1}{2} \hbar$.

HYPERCHARGE

A quality associated with some of the hadrons, similar to electric charge in that it occurs with both signs and its value for several particles is the sum of the individual values. Hypercharge is conserved in strong and in electromagnetic interactions, but not in weak interactions.

INTERACTION

A concept in quantum physics that replaces the notion of force in Newtonian physics. An interaction occurs whenever there is a change in the number, type, spins, or momenta of one or more particles that are near one another. Several kinds of interactions among subatomic particles have been identified, with varying properties.

ION

An atom in which the number of electrons is less than or more than the number of protons in the nucleus, giving the atom a net electric charge. Atoms in liquid solutions often become ions through transfer of electrons to other atoms.

ISOSPIN SYMMETRY

A mathematical relation among the properties of certain groups of hadrons. The particles that are related by isospin symmetry have the same spin, baryon number, and hypercharge and approximately equal rest energy, but have different electric charge. An example of particles related by isospin symmetry are the three types of pions.

ISOTOPES

Several nuclei containing the same number of protons but different numbers of neutrons. Isotopes have different rest mass and usually different nuclear properties, but atoms containing different isotopes have the same chemical properties. Many of the chemical elements have several stable isotopes and more unstable ones.

KINETIC ENERGY

A quality associated with moving objects whose amount is proportional to the mass of the object, and, in Newtonian physics, to the square of the speed of the object.

LEPTON

One of a group of subatomic particles characterized by having spin ½ $\hbar$ and no strong interactions. The group includes electrons, muons, two kinds of neutrino, and the antiparticles of each of these. Leptons do undergo weak and, in some cases, electromagnetic interactions.

LINEAR ACCELERATOR

A machine to accelerate particles to high energy. The accelerator is a long, straight tube, in which the particles move and are accelerated by electric forces to higher and higher energies.

METASTABLE

A system that decays spontaneously is said to be metastable if its half-life is much longer than other similar decaying systems. The subatomic particles that decay by weak interactions are examples of metastable particles.

MOLECULE

A combination of two or more atoms held together by the electrical forces acting between the electrons and nuclei in different atoms. Molecules range in size from H_2, a combination of two hydrogen atoms, to DNA, a combination of millions of atoms of various types.

MOMENTUM

A quality associated with moving objects. Its direction is that in which the object is moving, and its amount is, in Newtonian physics, proportional to the speed and to the mass of the object. This quality is sometimes called linear momentum to distinguish it from angular momentum.

MULTIPLET

A group of hadrons with similar properties, related by some symmetry such as isospin symmetry. The number of different hadrons in the multiplet is determined by the mathematical properties of the symmetry. The neutron and proton are a multiplet (doublet) for isospin symmetry.

MUON

An unstable particle with a mass about two hundred times greater than the electron but with the same electric charge and spin and many other similar properties. Because they decay with a half-life of about 10^{-6} sec, muons are not found in ordinary matter and instead are mainly produced in the decay of charged pions.

NEUTRINO

One of several types of particle which have zero rest mass, no electric charge, and spin $\frac{1}{2} \hbar$. An important property of neutrinos is that they interact only weakly, and therefore are difficult to detect, and can penetrate immense distances, even through dense matter.

NEUTRON

One of the two types of particle composing all nuclei. Neutrons have zero electric charge, spin ½ $\hbar$, and a rest energy of 940 MeV. Free neutrons are unstable, decaying into a proton, electron, and antineutrino.

NUCLEON

Either of the two particles—the proton and the neutron—found in ordinary nuclei. Since the properties, other than electric charge, of these two particles are very similar, they are often regarded as two aspects of the same kind of particle, called the nucleon.

PARITY SYMMETRY (P SYMMETRY)

A principle of symmetry implying an equal probability of occurrence for any two processes related by reversing the direction of the linear momentum of all particles involved in the process. Parity symmetry is satisfied in strong and electromagnetic interactions but not in weak interactions.

PARTICLE

An object with a definite mass, spin, and electric charge, whose size and internal structure are negligible, at least for some purposes. In present-day physics, atoms are not particles, because of their structure, while electrons are particles. However, in 1850, atoms could be treated as particles, and in 2050, electrons may not be considered particles.

PHOTON

The particles that compose a light ray. Photons have zero rest mass and travel at the speed of light. They carry 1 $\hbar$ unit of spin and are readily emitted or absorbed by charged particles.

PIONS

One of three types of meson, with charges of $+e$, $-e$, or zero, and rest energy of approximately 140 MeV. Pions are the lightest type of hadron and are copiously produced in high-energy collisions among nucleons. Each type is unstable, and decays into less massive particles.

POSITRON

The antiparticle of the elcctron, with a charge of $+e$. Positrons are not ordinarily present in matter on Earth, but can easily be made, together with electrons, by high-energy photons. Isolated positrons are stable, but an electron and positron can annihilate into photons.

POTENTIAL ENERGY

A quality associated with several objects, whose amount depends on the mutual distances of the objects. The potential and kinetic energy added together satisfy the law of conservation of energy (q.v.)

PROTON

Protons are one of the particles constituting all nuclei. They have a charge of $+e$, spin $\frac{1}{2}\,\hbar$, and rest energy of 938 MeV. Protons are the lightest baryons and so isolated protons do not decay. It is believed that most of the matter in the Universe is in the form of protons.

QUANTUM NUMBERS

The set of numerical values describing a quantum-mechanical state are called quantum numbers. For a single particle, the quantum numbers might be its charge, total angular momentum, spin projection, etc., or the values of any other complete set of quantities that can be determined together.

QUARK

One of several types of hypothetical subatomic particles. According to one model of hadrons, mesons are composed of a quark and an antiquark bound together while baryons are composed of three quarks. Quarks have not been observed as isolated objects, and according to some theories, cannot be.

RADIOACTIVITY

A behavior of certain types of atoms in which the nucleus undergoes a spontaneous transformation, emitting one of several types of high-energy particles. It is through the detection of these particles that radioactivity can be studied.

RELATIVITY PRINCIPLE

The basic idea of Einstein's special relativity theory, the statement that the laws of physics are the same for various observers who are moving at constant speed in a straight line with respect to each other. This principle places many restrictions on the laws of physics which are generally known as the requirements of relativistic invariance.

REST ENERGY

The energy of a particle when it is not moving and is far enough from other objects to be uninfluenced by them. According to special relativity theory, the rest energy is equal to the mass multiplied by the square of the speed of light.

SCATTERING

When atoms or subatomic particles hit one another, they may be deflected in their motion, transformed into other particles, or captured into bound states. All of these effects are called scattering processes or sometimes collision processes. If the outgoing objects are the same as the incoming ones, the scattering is called elastic; if not, it is called inelastic.

SCHRÖDINGER EQUATION

An equation that describes how the state of one or more particles changes with time. This equation plays the role in quantum mechanics that Newton's laws of motion do in Newtonian mechanics.

SPIN

A kind of angular momentum carried by some particles even when their velocity is zero. It can be pictured as a rotation of the particle about an axis passing through it. The spin of any particle or system of particles is either zero, a half-integer multiple of $\hbar$, or an integer multiple of $\hbar$.

STATE

In quantum mechanics, a condition of a particle or set of particles in which a maximum number of observable quantities are determined. For example, a single spinless particle in empty space with a definite momentum is in a state. Because of Heisenberg's relation, a state does not correspond to definite values for all observable quantities, and, in the example given, the position of the particle is completely undetermined.

STRONG INTERACTIONS

The set of processes by which the hadrons exchange energy and momentum and transform into one another. The adjective strong refers to the fact that these transformations occur with large probability whenever hadrons are near one another.

SYNCHROTRON

A circular machine that accelerates subatomic particles to high energy by the repeated action of electric forces on the particles. The particles are made to move in constant circular orbits by magnetic forces that continually increase in magnitude.

TACHYON

A hypothetical particle that always travels at a speed greater than that of light. The existence of tachyons is allowed, but not required, by the laws of quantum mechanics and special relativity.

TRANSITION

In quantum mechanics, a change of the state of one or more interacting objects. The change may involve the motion of the objects, or the type of objects involved. Individual transitions occur randomly, but the probability of various possible transitions can be predicted.

UNITARY SYMMETRY (SU3 SYMMETRY)

A mathematical relation among the properties of groups of hadrons that is more comprehensive than isospin symmetry. The particles related by unitary symmetry have the same spin and baryon number and roughly equal mass, but may have different hypercharge and electric charge. An example of a group of hadrons related by unitary symmetry is the eight, spin ½ $\hbar$, stable and metastable baryons.

UNSTABLE PARTICLE

A subatomic particle that spontaneously transforms (decays) into other subatomic particles is said to be unstable. If the half-life for the decay is relatively long, an unstable particle is called metastable.

VIRTUAL TRANSITION

A change from one state to another which lasts for only a very short time. Virtual transitions need not conserve energy because of the Heisenberg relation for energy and time, although they must satisfy the other conservation laws. The possibility of virtual transitions, especially those involving the creation and destruction of particles, has a major effect on the properties of the hadrons.

WAVE FUNCTION

A quantity that is associated with the motion of a particle, as described by quantum mechanics. The wave function varies from point to point in space and the square of its value at each point measures the probability of finding the particle at that point.

WAVELENGTH

For a wave that repeats its form periodically, the wavelength is the distance between two successive points at which the wave has the same intensity. The wavelength multiplied by the frequency is equal to the speed of the wave.

WEAK INTERACTIONS

A set of processes involving all types of subatomic particles by which particles transform into each other. The probabilities of such transformations are relatively small, which is why these interactions are called weak. Many of the symmetry properties and conservation laws satisfied in the strong interactions (q.v.), fail to be satisfied in weak-interaction processes.

Suggestions for Further Reading

I have listed here a few books for readers who wish to learn more about various topics in modern physics.

First a few books that describe some of the history of modern physics.

1. Boorse, H., and Motz, L. *The World of the Atom*. New York: Basic Books, 1966.
2. d'Abro, A. *The Evolution of Scientific Thought*. New York: Dover, 1950.
3. Livingston, M. S. *Particle Accelerators, A Brief History*. Cambridge, Mass.: Harvard University Press, 1969.

Some books that treat aspects of modern physics discussed in this book, using more mathematics than I do.

1. Frauenfelder, H., and Henley, E. M. *Subatomic Physics*. Englewood Cliffs, N.J.: Prentice-Hall, 1974.
2. Heisenberg, W. *The Physical Principles of the Quantum Theory*. New York: Dover, 1930.
3. Lapp, R. E., and Andrews, H. L. *Nuclear Radiation Physics*. Englewood Cliffs, N.J.: Prentice-Hall, 1972.
4. Taylor, E. F., and Wheeler, J. A. *Spacetime Physics*. San Francisco: W. H. Freeman, 1966.

Several topics in twentieth century physics are not treated in this book, because they are off the main direction. The reader can learn about some of them in the following books.

1. Holden, A. *The Nature of Solids*. New York: Columbia University Press, 1965.
2. Mendelsohn, K. *The Quest for Absolute Zero*. London: World University Library, 1966.

3. Sciama, D. *Modern Cosmology.* Cambridge, England: Cambridge University Press, 1971.

A brilliant, though idiosyncratic introduction to physics, on a relatively advanced level is given in these books.

Feynman, R., *et al. Feynman Lectures in Physics,* Vols. 1, 2, 3. Reading, Mass.: Addison-Wesley, 1963.

Index

WHAT IS THE WORLD MADE OF?

GERALD FEINBERG

Twentieth-century science has sought to understand and describe the ultimate composition of matter. This quest has led physicists beyond the atom to the realm of subatomic particles and other phenomena whose discovery has revolutionized our understanding of the basic components of the world.

Seeking to make the concepts which explain the nature of these particles accessible to non-scientists, Gerald Feinberg has written a book that elucidates the achievements of twentieth-century physics. He begins with a concise explanation of quantum theory, and traces the search for the fundamental building blocks of our world from atoms through electrons and nuclei to the currently investigated realm of hadrons, leptons, and quarks. His non-mathematical approach provides a lucid, yet in-depth explanation of subatomic phenomena and the laws